Design for Bending, Torsion and Buckling

A monograph covering

- Bending stresses in beams with symmetric and non-symmetric cross sections
- Shear stresses and twist in solid and hollow circular and non-circular cross section shafts
- Critical buckling loads, design criteria and axial compression of slender members from lateral instability

Carl F. Zorowski

Design Engineering Monograph V

Design for Bending, Torsion and Buckling

Design for Bending

Design for Torsion

Design for Buckling

Preface

Many machine parts possess an axial dimension larger than those that define the remainder of its geometry. If loads are applied laterally to this axial direction the part will undergo bending. This gives rise to a characteristic internal normal stress distribution. If the part is subjected to an applied torque about this axis the element will twist. This creates a state of internal shear stress. If the length of the element is large compared to its cross sectional dimensions an axial compressive force can produce a lateral bending instability. This can occur at stress levels below those required to produce material failure.

This monograph deals with predicting the internal stress states created by these three kinds of deformation behavior. In addition to presenting classical solution models applicable to simple geometries and loadings the content is expanded to cover more complex circumstances not always covered in standard texts on the subject.

Chapter 1 of Design for Bending begins with the classic "simple beam theory" model. Based on the assumptions that plane cross-sections remain plane during bending and the material obeys Hooke's Law this model predicts a linear internal normal bending stress distribution that balances the bending moment at that location. However, it is only applicable to a beam whose cross section possesses an axis of symmetry. This model is expanded and generalized to apply to beams possessing any general cross section geometry with no restriction on how the loading is applied.

In Chapter 2 advantage of the area properties about rotated axes in the cross-section permit a simplification of the general stress formulation. This leads to a convenient method of determining the neutral axis of bending and the location and magnitude of the maximum stress.

Two numerical examples are presented in Chapter 3 that demonstrate the application of this procedure to specific right angle and z cross section beams. The results are compared to stress values calculated using the simple beam model formula in Chapter 1.

Chapter 1 of Design for Torsion begins with the classic model for determining the maximum shear stress and unit twist in a solid or hollow circular shaft subjected to an applied torque. This is followed by the theory of elasticity prescription for analyzing solid noncircular cross section shafts. An approximate solution to this classic formulation is presented by analysis of a pressurized membrane as an analogy to the theory of elasticity formulation. This gives the same results as the simple circular shaft model. The membrane analogy is then used to solve the torsion of a thin rectangular cross section shaft with the results compared to the theory of elasticity solution of twisted rectangular cross sections. The approximate thin rectangular torsion solution is adapted to solve numerically the torsional behavior and induced stresses of an extruded H section shaft.

In Chapter 2 the membrane analogy is used to solve the problem of a hollow non-circular cross section shaft. This gives rise to several simple formulae in which the wall thickness is constant. This formulation is used to numerically compare this approximation to the exact solution for a thin walled circular shaft from the simple model. Comparisons are included of different geometries that contain the same amount of material, possess the same cross section circumferential length or are slit along the length of the shaft. A general solution for a two-cell hollow system is formulated and solved for the internal stresses. This solution is applied to a circular hollow section with a diametral web to illustrate the effect of this change in geometry.

Chapter 3 deals with the numerical solution of a practical design problem in which a convoluted thin walled cross section tube is proposed to provide greater torsional flexibility in a gas tight connection between two tanks. The results are compared with a thin walled circular tube with the same outside diameter.

Chapter 1 of Design for Buckling solves the problem of the lateral bending of a slender fixed-free column subjected to an axial compressive load applied with eccentricity to the centroid of its cross section. This leads to the classic Euler Buckling Load as the eccentricity approaches zero defining the maximum compressive load that can be applied before undefined lateral instability occurs.

In Chapter 2 the Euler buckling load is determined for the three remaining classic column support conditions i.e. pined-pined, pined- fixed and fixed-fixed end restraints. These are all summarized in one formulation multiplied by a constant dependent on the column support.

Chapter 3 deals with the effects of cross sectional geometry and material properties on the buckling behavior. This is accomplished by rearranging the Euler buckling load into a formulation that permits establishing a design criterion between short and long columns dependent on modulus of elasticity and moment of inertia of the cross section. Numerical examples for these effects are demonstrated for steel and aluminum.

When columns undergo lateral instability deflection the ends move closer to one another. Chapter 4 develops a general formula for calculating the amount of this movement in terms of the maximum lateral displacement and the deflected shape of the column. The lateral bending behavior of a thin plastic ruler in a desktop buckling experiment is used to illustrate the significance and numerical magnitude of this phenomenon.

In Chapter 5 the materials from the previous four chapters are used to numerically solve the redesign of an existing temperature activated switch. Its operation involves both buckling and lateral displacement behavior produced by a specific temperature change.

As in previous monographs about one third of the content of this publication appears as figures that present mathematical developments, illustrations and graphical materials. The level of mathematics is somewhat higher than in previous monographs. This is required by the greater complexity of the models employed and the principles of mechanics involved. It is strongly recommended that the reader follow through with the details of the mathematical developments to better understand and appreciate the results being sought as well as the elegance of the mathematics. The text provides assistance to help the reader work through the more involved mathematical manipulations.

The materials contained in this monograph are from notes used in a Mechanical Design Engineering course taught at North Carolina State University by the author. An audio supplemented version of the content is available at *http://www.designengineeringreview.com*.

Carl F. Zorowski
December 2016
Cary, NC

Design for Bending

Chapter 1 Bending Stress In Beams

Chapter 2 Bending about Neutral Axis

Chapter 3 Example Problems

Design for Bending, Torsion and Buckling

Chapter 1 Bending Stresses in Beams

The bending portion of this monograph begins with the calculation of stresses in bent beams using "simple beam theory". This only applies to beam cross-sections that possess an axis of symmetry. The remainder of the content deals with determining bending stresses in beams that do not possess an axis of symmetry, a more complex process.

Classic Bending Behavior

When structural members that are long compared to their transverse dimensions are subjected to lateral loading they under go transverse deflections as depicted in Figure 1-1. In this example the top of the beam is stretched while the bottom is compressed. This results in tension stresses in the upper portion and compressive stress in the lower part.

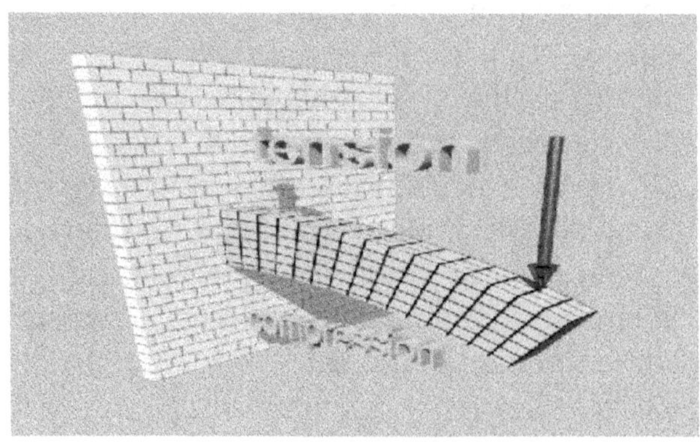

Figure 1-1 Classic Bending Behavior

Calculating the magnitude of these stresses and their distribution across the depth of the beam is the subject of bending theory.

Simple Beam Theory Model

If lateral deflections are small compared to the beam length the deformation model that adequately describes this behavior is that plane cross sections of the beam prior to bending remain plane after deflection occurs. This model leads to:

1. An assumed linear strain distribution across the depth of the beam that results in a linear stress distribution assuming hook's law is valid for the material.
2. Application of equilibrium requires the resultant of this internal stress distribution to equal the magnitude of the local bending moment.
3. Satisfying the stress resultant/local bending moment requirement results in the transition from tension to compressive stress through the centroid of the cross section.
4. If the cross section of the beam possesses an axis of symmetry the bending stress is given by a simple expression relating the bending moment and the cross sectional area moment of inertia.

Simple Beam Stresses

Shown in Figure 1-2 is a cantilever beam subjected to a bending moment sector M_x about the x axis.

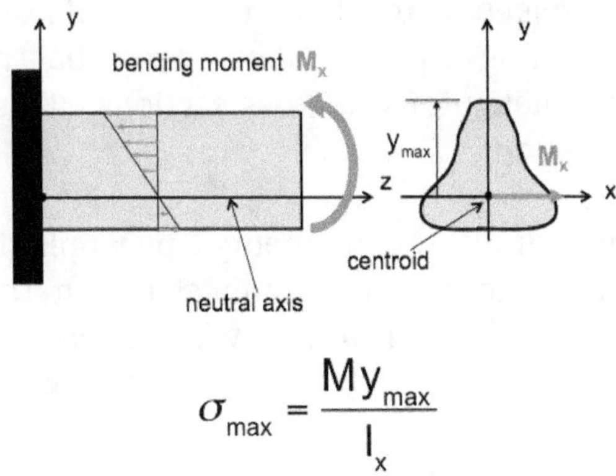

$$\sigma_{max} = \frac{My_{max}}{I_x}$$

Figure 1-2 Simple Beam Theory

This will give rise to compressive stresses in the upper portion of the beam and tension in the lower portion. The distribution of the stress is linear based on the assumptions that planes sections remain plane after bending and hook's law is applicable. The point where the stress transitions from compression to tension passes through the centroid of the section. This is taken as the origin of the coordinate system. Bending about the x axis, an axis perpendicular to the axis of symmetry, gives rise to a linear distribution of normal stress in the yz plane with the maximum stress given by the relationship, σ_{max} is equal to the product of the bending moment M_x and the furthest distance, y_{max}, from the x axis divide by the area moment of inertia about the x axis I_x, i.e. $\sigma_{max} = M_x\ y_{max}/I_x$.

Unsymmetrical Bending

Shown in Figure 1-3 are two beam cross sections subjected to bending moments M_o that produce-bending stresses about axes that are not axes of symmetry for the cross sections. This raises several questions:

1. Can "simple beam theory" be applied to determine the maximum stress in these cross sections, if not how is the maximum stress determined?

2. How much difference is there between what simple beam theory would predict and the actual value of the maximum stress?

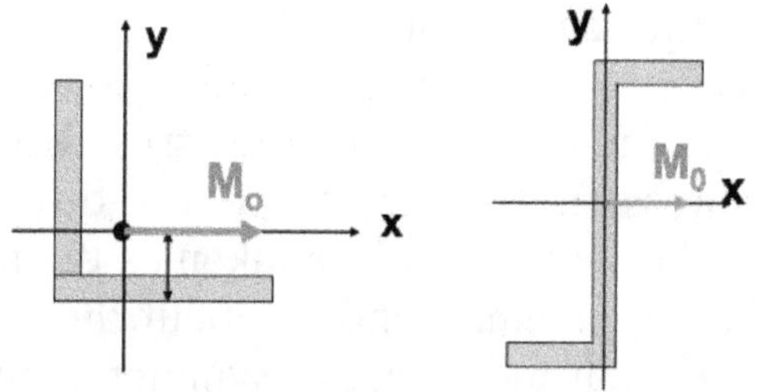

Figure 1-3 Unsymmetrical Cross Section

Generic Stress Equation

Consider the bending of a beam of some general cross section by bending moment components about two perpendicular axes as shown in Figure 1-4. The vector M_x denotes bending about the x axis and the vector M_y denotes bending about the y axis. Both of these bending moments will produce a normal stress, σ, in the z direction on the incremental element of area dA of the cross section. To determine how that stress is related to the applied moments it is assumed that the distribution of normal stress will vary linearly with the coordinates x and y. This is expressed as the equation: σ is equal to $C_1 + C_2\,x + C_3\,y$. The task now is to determine how these constants are related to the applied moments and the area properties of the cross section.

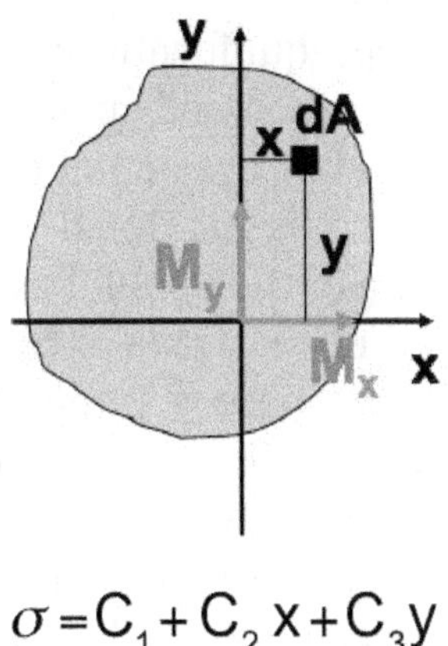

$$\sigma = C_1 + C_2\,x + C_3\,y$$

Figure 1-4 Bending about Two Axes

Satisfying Equilibrium

As in "simple beam theory" the force and moment resultants of the stress distribution must be equal to the applied loading at that section. This requires applying and satisfying equilibrium with the assumed linear general expression for the stress distribution. Assuming in the most general case that an axial load is also applied on the beam then three conditions of equilibrium must be satisfied. The first is that the net resultant force due to the stress distribution is equal to the axial force. The second is the net moment of the stress distribution about the x-axis is equal to M_x and finally that the next resultant moment of the stress distribution about the y axis is equal to M_y.

Equilibrium Equations

There are three equilibrium conditions to be satisfied. These are expressed mathematically in Figure 1-5. The normal stress acting on the area dA integrated over the cross sectional area must be equal to the magnitude of the applied axial force, P, in the z direction. The normal stress times dA multiplied by the coordinate y integrated over the cross section must be equal to the magnitude of the bending moment M_x at this section. The normal stress times dA multiplied by the coordinate x integrated over the cross section must be equal to the magnitude of the bending moment M_y at the section.

A negative sign is introduced into the third equation to insure a tensile stress for a positive coordinate and positive moment. These three equations will permit the determination of the three constants in the assumed linear stress distribution

Axial force on cross section

$$P = \int_A \sigma dA$$

Moment component about the x axis

$$M_x = \int_A \sigma y dA$$

Moment component about the y axis

$$-M_y = \int_A \sigma x dA \quad \text{negative to account for } +M_y \text{ direction}$$

Figure 1-5 Equilibrium Equations

Axial Load

Following substitution of the general stress equation into the axial force equilibrium equation can be interpreted as three separate integrals multiplied by the constants, C_1, C_2 and C_3. The integral of dA over the cross section is simply the area of the cross section, A. If the origin of the xy axis system is taken through the centroid of the area then the integral of xdA over the cross section is zero as this first moment is really how the location of the centroid in defined. In a similar fashion the integral of ydA over the area will also be zero.

The final result is that the constant C_1 is simply equal to P/A. This is a uniform stress due a tensile (or compressive) force P applied to the beam in the z direction (Figure 1-6).

$$P = \int \sigma dA = \int (C_1 + C_2 x + C_3 y) dA$$
$$P = C_1 \int dA + C_2 \int x dA + C_3 \int y dA$$

but $\int dA = A$, and $\int x dA = 0$, $\int y dA = 0$

if origin of axes passes through centroid of cross section

then $C_1 = \frac{P}{A}$

Uniform stress due to tensile force P

Figure 1-6 Axial Load Equation

M_x Moment

Substituting the assumed stress equation in to the M_x equilibrium condition in Figure 1-5 again leads to interpreting three separate integrals. The constant C_1 is multiplied by the integral of ydA over the area. This integral is zero because the origin of the xy axes system is at the centroid of the area. The integral of xydA is nothing more the second area property designated as the product of inertia I_{xy}. The final integral of y^2dA is the second moment of the area designated as the moment of inertia I_x.

This results in an equation for M_x involving the constants C_2 and C_3 along with the second moment properties of the area I_{xy}, the product of inertia, and I_x, the moment of inertia with respect to the x axis (see Figure 1-7).

$$M_x = \int \sigma y\,dA = \int (C_1 + C_2 x + C_3 y)\,yd$$
$$M_x = \int (C_1 y + C_2 xy + C_3 y^2)\,dA$$
$$M_x = C_1 \int y\,dA + C_2 \int xy\,dA + C_3 \int y^2\,dA$$
$$\text{but } \int y\,dA = 0, \text{ and } \int xy\,dA = I_{xy}, \int y^2\,dA = I_x$$

moment and product of inertia about x and y axes

$$\therefore M_x = C_2 I_{xy} + C_3 I_x$$

Figure 1-7 M_x Moment Equation

M_y Moment

The third and final equation of equilibrium to be satisfied is that the integral of σ times xdA over the cross section must be equal to the moment component M_y. A negative sign is introduced so that tension stresses are always positive whether created by M_x or M_y. Again three integrals need to be interpreted. The integral of xdA over the area is zero since the coordinates axes pass through the centroid of the cross section. The integral of x^2dA over the cross section is simply the moment of inertia about the y-axis, I_y, while the integral of xydA is again the product of inertia of the cross section, I_{xy}.

This results in a second equation involving the constants C_2 and C_3 together with the moment component M_y and the second moment area properties of the cross section (see Figure 1- 8).

$$-M_y = \int \sigma x\, dA = \int (C_1 + C_2 x + C_3 y) x dA$$
$$-M_y = \int (C_1 x + C_2 x^2 + C_3 xy) dA$$
$$-M_y = C_1 \int x\, dA + C_2 \int x^2\, dA + C_3 \int xy\, dA$$
$$\text{but } \int x\, dA = 0, \text{ and } \int x^2\, dA = I_y, \ \int xy\ dA = I_{xy}$$

moment and product of inertia about x and y axes

$$\therefore -M_y = C_2 I_y + C_3 I_{xy}$$

Figure 1-8 M_y Moment Equation

Simultaneous Equations

The constant C_2 and C_3 can now be determined from the two equilibrium equations for M_x and M_y in Figure 1-7 and Figure 1-8. Solving the equations simultaneously in Figure 1-9 results in expressions for the constants that involve both the bending moment components and the cross section area properties. It is observed that the denominators of these solutions for C_2 and C_3 are identical.

Two equations for C2 and C3

$$M_x = C_2 I_{xy} + C_3 I_x$$
$$-M_y = C_2 I_y + C_3 I_{xy}$$

Solve simultaneously to give

$$C_2 = -\frac{M_y I_x + M_x I_{xy}}{(I_y I_x - I_{xy}^2)}, \quad C_3 = +\frac{M_x I_y + M_y I_{xy}}{(I_y I_x - I_{xy}^2)}$$

Figure 1-9 Constants C_2 and C_3

Also the numerators have the same format of the moment component times the respective cross section moment of inertia plus the other moment component times the cross section product of inertia. The negative sign is again a consequence of having positive moment component always produce positive tensile stresses.

General Stress Equation

The final general equation for the stresses in a beam of any cross section bent by two bending moment components and subjected to an axial load is given in Figure 1-10.

$$\sigma = \frac{P}{A} - \frac{\left(M_y I_x + M_x I_{xy}\right)}{\left(I_y I_x - I_{xy}^2\right)} x + \frac{\left(M_x I_y + M_y I_{xy}\right)}{\left(I_y I_x - I_{xy}^2\right)} y$$

Figure 1-10 General Stress Equation

Consider the special case of loading in Figure 1-11 in which there is no axial load and bending only occurs about the x-axis in a rectangular cross section beam.

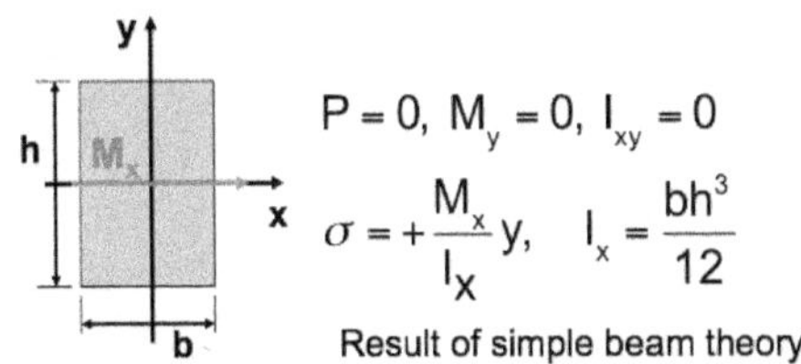

Figure 1-11 Special Case Loading

P is equal to zero, My is zero and because the rectangle is symmetric with respect to the x-axis then I_{xy} the product of inertia is zero. Substituting these conditions into the general bending stress equation results in $\sigma = M_x\ y\ /\ I_x$ which is the result given previously from "simple beam theory".

Chapter 2 Bending about Neutral Axis

Another more convenient formulation of the general bending stress equation can be obtained following a consideration of how the area properties of the cross section change with the orientation of the xy axis system through its centroid.

Rotated Axes Formulation

This is accomplished by computing the moments of inertia and product of inertia with respect to a rotated axis system x'y' relative to the xy axis system. This requires determining the coordinates x' and y' in terms of x, y and the angle of rotation, θ, relative to the incremental area dA in the cross section.

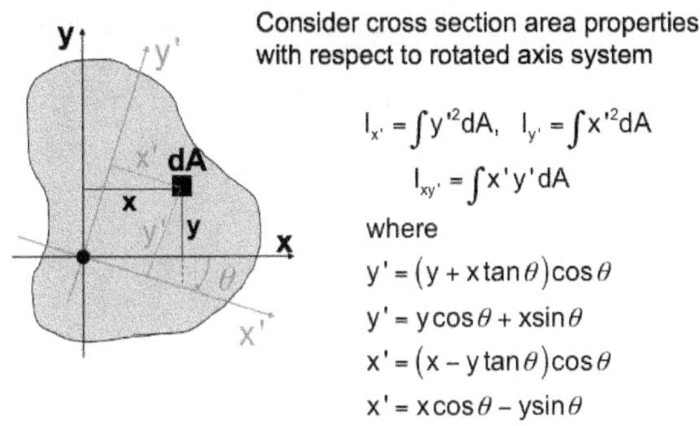

Figure 2-1 General Rotated Axes

On the right side of the Figure 2-1 are the defining integrals for $I_{x'}$, $I_{y'}$ and $I_{x'y'}$ along with the relationships for the coordinates x' and y' to the incremental area dA in terms of the coordinate x and y to dA and the angle of rotation, θ.

Moments of Inertias about x' and y'

$I_{x'}$ is determined by integrating the expression y'^2dA over the cross section with y' replaced by its equivalent expression in terms of x, y and θ from Figure 2-1. This results in three integrals in Figure 2-2 that can be interpreted in terms of area properties with respect to the xy axis system. The result is that $I_{x'}$ is equal to $I_x \cos^2\theta + I_y \sin^2\theta + I_{xy}(2\sin\theta\cos\theta)$. $I_{y'}$ is evaluated in a similar fashion in Figure 2-2 to give the final result that I_y' is equal to $I_y \cos^2\theta + I_x \sin^2\theta - I_{xy}(2\cos\theta\sin\theta)$.

$$I_{x'} = \int y'^2 dA = \int (y\cos\theta + x\sin\theta)^2 dA$$

$$I_{x'} = \cos^2\theta \int y^2 dA + 2\cos\theta\sin\theta \int xydA + \sin^2\theta \int x^2 dA$$

$$\text{but } I_x = \int y^2 dA, \quad I_y = \int x^2 dA \text{ and } I_{xy} = \int xydA$$

$$I_{x'} = I_x \cos^2\theta + I_y \sin^2\theta + I_{xy} 2\cos\theta\sin\theta$$

$$I_{y'} = \int x'^2 dA = \int (x\cos\theta - y\sin\theta)^2 dA$$

$$I_{y'} = \cos^2\theta \int x^2 dA - 2\cos\theta\sin\theta \int xydA + \sin^2\theta \int y^2 dA$$

$$\text{but } I_y = \int x^2 dA, \ I_x = \int y^2 dA \text{ and } I_{xy} = \int xydA$$

$$I_{y'} = I_y \cos^2\theta + I_x \sin^2\theta - I_{xy} 2\cos\theta\sin\theta$$

Figure 2-2 Ix' and Iy' Relative to Rotated Axes

Product of Inertia bout x'y' axes

The product of inertia $I_{x'y}'$ is defined as the integral of x'y'dA over the cross section. Again x' and y' are replaced by their equivalent expressions in x, y and θ before the integration is carried out. Interpreting the three resulting integrals in Figure 2-

3 in terms of properties with respect to the xy axis system gives a final result for $I_{x'y'}$ as the quantity:

$$(I_y - I_x) \sin\theta \cos\theta + I_{xy} (\cos^2\theta - \sin^2\theta)$$

$$I_{xy'} = \int x'y'dA = \int (y\cos\theta + x\sin\theta)(x\cos\theta - y\sin\theta)dA$$

$$I_{xy'} = (\cos^2\theta - \sin\theta^2)\int xy\,dA + \cos\theta\sin\theta\int x^2 dA - \cos\theta\sin\theta\int y^2 dA$$

$$\text{but but } I_{xy} = \int xy\,dA,\ I_x = \int y^2 dA \text{ and } I_y = \int x^2 dA$$

$$I_{xy'} = (I_y - I_x)\sin\theta\cos\theta + I_{xy}(\cos^2\theta - \sin^2\theta)$$

Figure 2-3 $I_{x'y'}$ Relative to Rotated Axes

Comparison with 2D Stresses

Listed at the top of Figure 2-4 are the three equations for $I_{x'}$, $I_{y'}$ and $I_{x'y'}$ for the rotated axes x'y' in terms of the angle of rotation, θ, and the properties I_x. I_y and I_{xy} relative to the xy axis system. If substitutions for normal and shear stress as listed are made in these equations they become the equations for a two-dimensional stress state with respect to a rotated axis system.

The importance of this is that just as a two-dimensional stress system can be graphically represented by a Mohr circle construction so too can the moments of inertia and product of inertia of a cross section be represented graphically in the same way.

$$I_{x'} = I_x \cos^2\theta + I_y \sin^2\theta + I_{xy} 2\cos\theta\sin\theta$$
$$I_{y'} = I_y \cos^2\theta + I_x \sin^2\theta - I_{xy} 2\cos\theta\sin\theta$$
$$I_{xy'} = \left(I_y - I_x\right)\sin\theta\cos\theta + I_{xy}\left(\cos^2\theta - \sin^2\theta\right)$$

If the following substitutions are made

$$I_{x'} \Rightarrow \sigma_{x'}, \quad I_{y'} \Rightarrow \sigma_{y'}, \quad I_{xy'} \Rightarrow \tau_{xy'}$$
$$I_x \Rightarrow \sigma_x, \quad I_y \Rightarrow \sigma_y, \quad I_{xy} \Rightarrow \tau_{xy}$$

The equations are similar to those for a two dimensional rotated stress state

Figure 2-4 Comparisons with 2D Stresses

Mohr Circle for Area Properties

If I_x, I_y and I_{xy} for a given cross section are known the space coordinates (I_x, I_{xy}) and (I_y, $-I_{xy}$) can be plotted on an axis system where moments of inertia are plotted on a horizontal axis and product of inertia are plotted on a vertical axis. Connecting these two space coordinate points by a straight line defines the diameter of the Mohr circle for area properties (see Figure 2-5).

As with stresses the diameter of the Mohr circle that coincides with the horizontal axis defines the maximum and minimum values of I_x and I_y. These are referred to as the principle moments of inertia I_1 and I_2

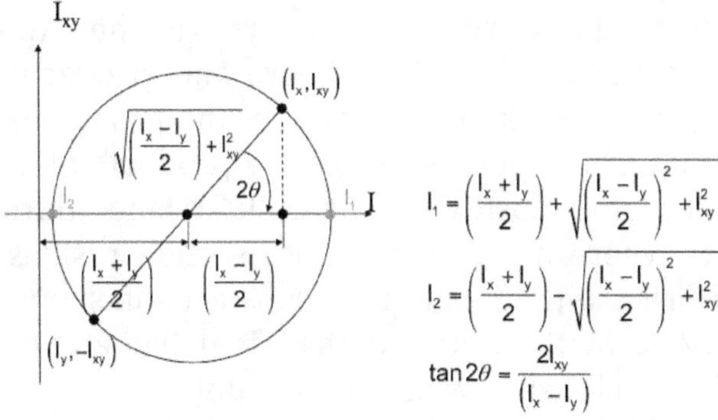

Figure 2-5 Mohr Circle for Area Properties

It is observed that for this particular orientation of axes on the cross section the associated product of inertia is zero. The magnitudes of the maximum and minimum moments of inertia in terms of Ix, Iy and Ixy are listed on the right side of Figure 2-5. Note their similarity to the equations for maximum and minimum values of normal stress in a two dimensional state.

The angle of rotation from the original xy axis system to the axes of the principal moments of inertia is given by the relation that $\tan 2\theta = 2I_{xy} / I_x - I_y$. Rotation of the axes on the cross section is in the same direction as on the Mohr circle.

Application to Z Section

The Mohr circle construction for area properties is now applied to a generic Z cross-section bent by the moment M_0 about the x-axis. Since most of the area of the cross section is located in the first and third quadrants of the xy axis system and it is further from the x axis than the y axis then the moment of inertia I_x will be positive and greater in value than I_y. By a similar argument the product of inertia will also be positive since the products of xy coordinates are positive in the first and third quadrants where most of the cross section resides. This gives rise to the Mohr circle construction shown in Figure 2-6. With rotation of the initial diameter on the Mohr circle clockwise by amount 2θ to the diameter defining the principle moments of inertia the xy axis system must be rotated clockwise through the angle θ to establish the x'y' axis system about which the moments of inertia are maximum and minimum and the product of inertia is zero.

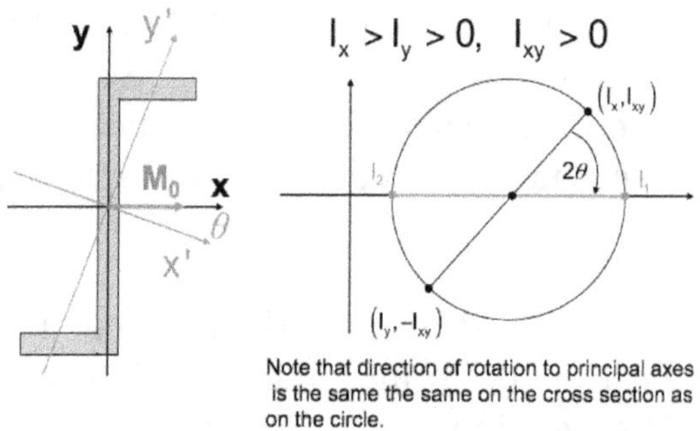

Figure 2-6 Properties of Generic Z Section

Stress Calculations

The general stress equation is now applied relative to the axis system defining the principle moments of inertia that are now designated as xy rotated through angle theta from the horizontal.

The original bending moment is also resolved into its components with respect to the xy axis system as $M_0\sin\theta$ and $M_0\cos\theta$. Substituting the moment components and area properties measured relative to the xy principle axis system into the general stress equation results in the simplified form at the bottom of Figure 2-7. The two terms in this equation resemble "simple beam theory" applied with respect to the principle axes of the cross section.

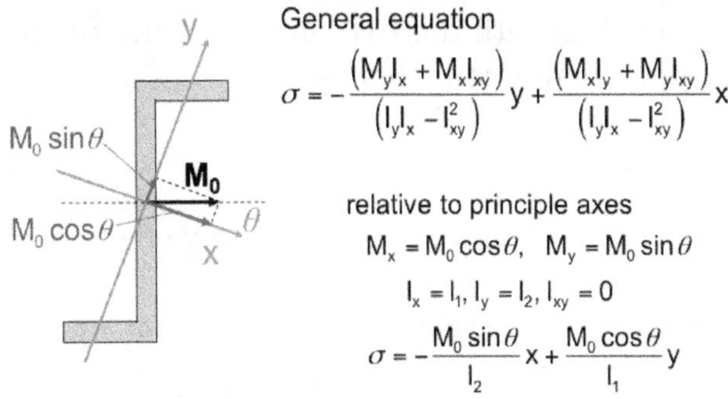

Figure 2-7 Rotated Axes Stress Equation

Maximum Stress

With this simpler form of determining the stress in unsymmetrical cross sections there still remains the question of the magnitude of the maximum stress and where is it located. By setting σ in the general stress equation relative to the principle axes equal to zero a linear equation is obtained which defines the neutral axis of the bending stress distribution through the centroid of the section.

The orientation of the neutral axis is obtained by differentiating this equation for y by x to give the tangent of alpha, α, the angle of the neutral axis measured positive counter clockwise from the x axis. The maximum stress will occur at that location in the cross section that is at the furthest distance measured perpendicular from the neutral axis as shown in Figure 2-8. The maximum stress is calculated by first determining the x and y coordinates of this point and substituting them into the simplified general stress equation together with the two moment components and the principle moments of inertia.

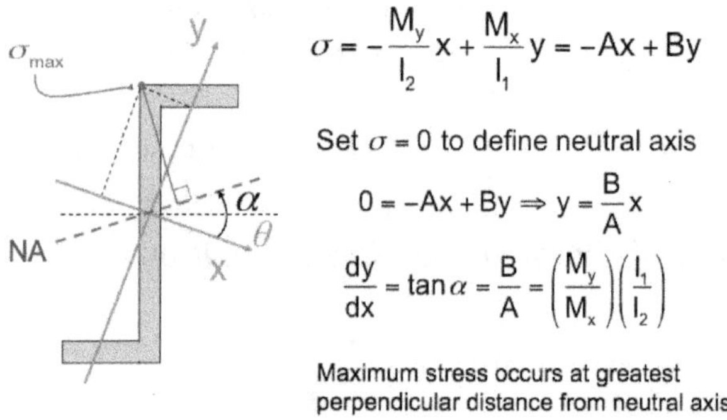

Figure 2-8 Calculation of Maximum Stress

Chapter 3 – Example Bending Problems

Chapter 3 of Design for Bending presents the solution to example problems in which the maximum bending stress in a right angle and Z section beam is determined using the theory in Chapter 1. Results are compared with the values obtained from application of "simple beam theory".

Example Problem 1 – Right Angle Section

In this example a 1 by 1 by 1/8 inch right angle section is bent by a moment M about the x axis as shown in Figure 3-1. This represents bending about an axis that is not an axis of symmetry even though the section possesses an axis of symmetry. The numerical values of I_x and I_y as well as the location of the centroid come from a Steel Sections Handbook. To determine the principle moments of inertia it is also necessary to know the value of I_{xy} that is not listed in the handbook.

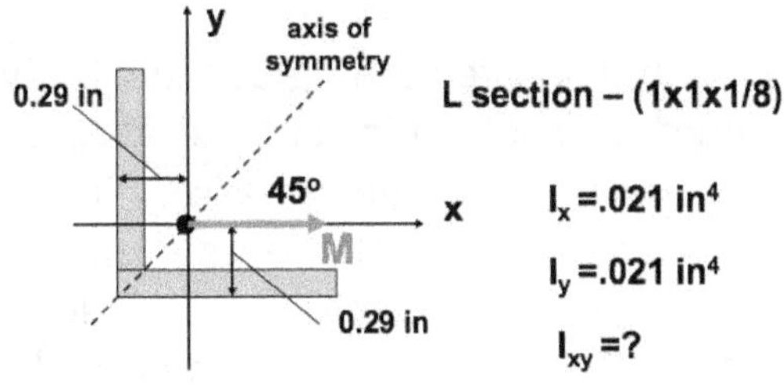

Figure 3-1 Right Angle Properties

Calculation of I_{xy}

To determine the product of inertia I_{xy} of the right angle section it is first divided into two rectangular areas A_1 and A_2 in Figure 3-2. I_{xy} of the angle is then calculated as the sum of products of inertia of A_1 and A_2 about heir respective centroid axes parallel to the xy axis system plus the product of their areas times the x and y distances from their respective centroids to the centroid of the angle section.

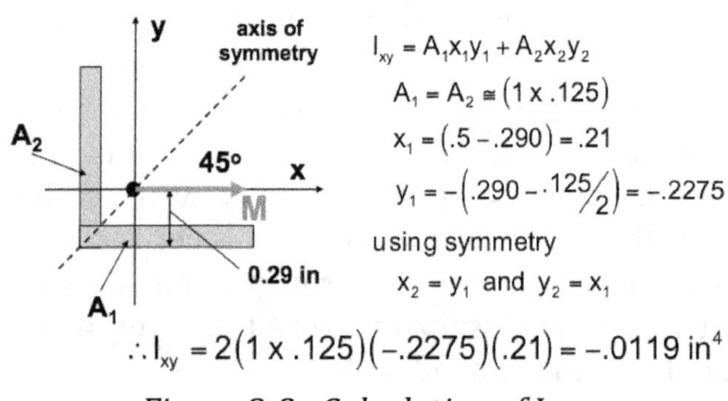

Figure 3-2 Calculation of I_{xy}

Since A_1 and A_2 are rectangles their individual products of inertia about their centroid axes parallel to the x and y axes of the angle are zero. This makes the product of inertia of the angle simply the areas A_1 and A_2 times their respective x and y distances from their centroids to the centroid of the angle. With the areas of A_1 and A_2 equal and the distances from their centroids to the centroid of the angle mirror images of one another I_{xy} is then 2 (1) (0.125) inches for the area times -0. .2275 and 0.21 as the distances from the centroids of A_1 and A_2 to the centroid of the angle.

The result is -0.0119 in^4. The negative value agrees with most of the material being in the second and fourth quadrants.

Principle Moments of Inertia

The Mohr's Circle in Figure 3-3 is now used to determine the magnitude of the principle moments of inertia and the angle of the principle axes. With I_x and I_y being equal the diameter that defines the circle is a vertical line through the center of the circle at 0.021 in^4. From the geometry of the circle I_1 is simply 0.021 plus 0.0119 or 0.0329 in^4.

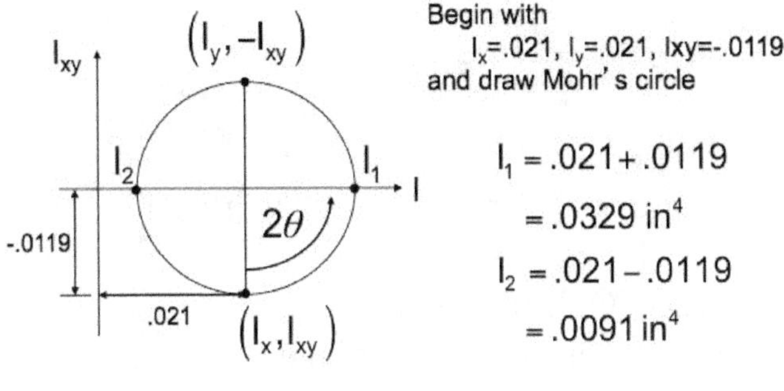

Figure 3-3 Mohr's Circle for Right Angle

I_2 is 0.021 minus 0.0119 or 0.0091 in^4. It is observed that 2θ is 90 degrees making θ =45 degrees. This is appropriate for an angle section whose two legs are the same length.

Stress Equation

Resolving the applied moment M into its two components along the x' and y' axes and substituting these into the general stress equation along with the values of I_1 and I_2 results in the final equation for the stress as $\sigma = (0.707\ M_o)\ ((x'/0.0091)\ +(y'/\ .0329))$ as shown in Figure 3-4.

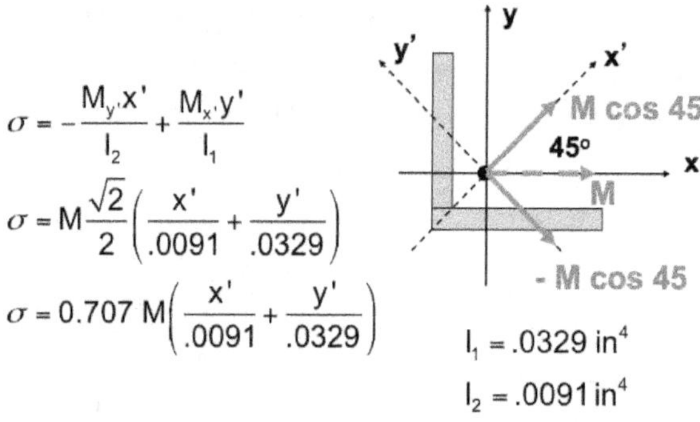

Figure 3-4 General Stress Equation

Neutral Axis Orientation

The orientation of the neutral axis is now determined by setting the equation for sigma, σ, equal to zero and performing the differentiation dy'/dx' to give the tangent of alpha, α. Numerically tan α = -3.61 which results in α being -74.5 degrees As shown on Figure 3-5 the neutral axis is the heavy dotted line rotated 74.5 degrees clockwise from the x' axis. The furthest perpendicular distance from the neutral axis where the stress will be maximum is at the top outer most point on the vertical leg of the angle.

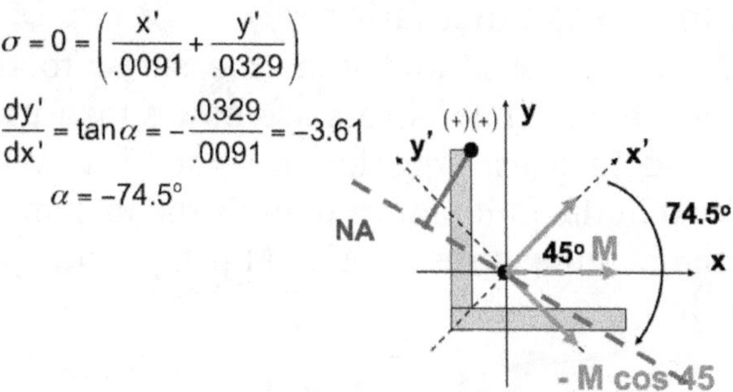

Figure 3-5 Location of Neutral Axis

Maximum Stress Location

The values of x' and y' to the point of maximum stress are calculated using two right triangles on Figure 3-6. The distance y' is determined from the light gray triangle whose hypotenuse is the height of the vertical leg. The distance x' is given by y'-a where "a" is the hypotenuse of a smaller lower triangle whose base and height are both (0.29-0.125). Approximate values of y' and x' are 0 .662 and 0.428 inches respectively.

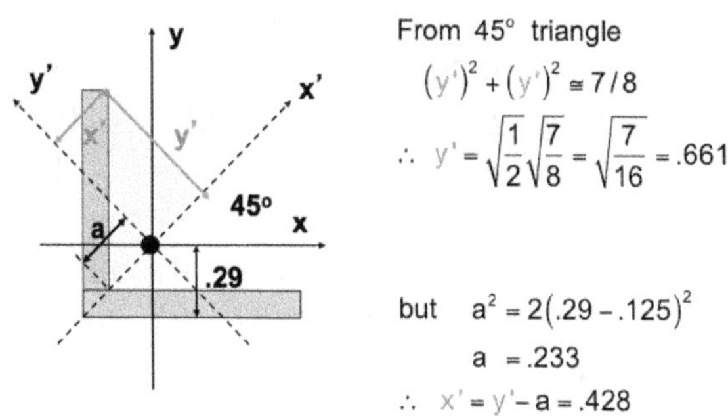

Figure 3-6 Coordinates to Maximum Stress

Maximum Stress Calculation

The values of x' and y' as measured to the location of the maximum stress are now substituted into the bending stress equation in Figure 3-4. With M expressed in the unit of inch pounds the maximum value of the bending stress is 47.5 M psi, pounds per square inch.

$$\sigma = 0.707\ \mathrm{M}\left(\frac{x'}{.0091} + \frac{y'}{.0329}\right)$$

substitue $x' = .428$ and $y' = .661$

$$\sigma = 0.707\ \mathrm{M}\left(\frac{.428}{.0091} + \frac{.661}{.0329}\right)$$

$$\sigma = 0.707\ \mathrm{M}(47.03 + 20.09)$$

$$\sigma = 47.5\ \mathrm{M}$$

Figure 3-7 Maximum Stress Calculation

Compare with Simple Beam Theory

If simple beam theory is used to calculate the maximum stress in the section it would be given simply by My_{max}/I In this case y_{max} would the vertical distance from the angle centroid to the outer most location or 1 - 0.29 inches and the moment of inertia I would be I_x relative to the x axis or 0.021 in^4. This gives a maximum value for the stress of 33.8 M. The percentage difference between the actual maximum stress and the approximate maximum stress is 29% with simple beam theory giving the lower value, see Figure 3-8.

This clearly illustrates the importance of properly calculating the maximum stress in sections that are not bent about axes of symmetry.

$$\sigma = \frac{M}{I}y = M\frac{(1-.29)}{.021} = 33.8M$$

$$\frac{\sigma_{exact} - \sigma_{approx}}{\sigma_{exact}} = \frac{47.5-33.8}{47.5} = .29$$

$$\sigma_{approx} \cong 29\ \%\ \text{low}$$

Figure 3-8 Comparison with Simple Beam Theory

Example Problem 2 –Z Section

In example problem 2 the Z section In Figure 3-9 is subjected to a bending moment, M_o, applied about a horizontal axis through its centroid. This represents general bending of a section that does not possess an axis of symmetry. The maximum stress will again be calculated and compared to that obtained by applying simple beam theory.

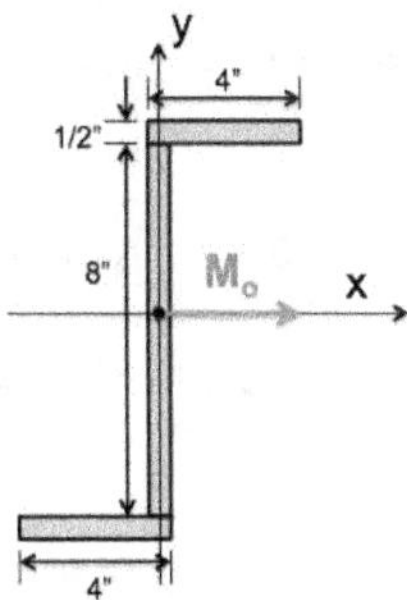

Figure 3-9 Z Section Dimensions

Calculate I_x, I_y and I_{xy}

To calculate I_x, I_y and I_{xy} for the Z section it is first separated into the rectangles, A_1 and A_2, (see Figure 3-10). I_x then becomes the moments of inertia of A_1 and A_2 about their respective centroids relative to axes parallel to the x-axis plus the areas of A_2 times the y distance between its centroid and that of the Z section squared. The moment of inertia of a rectangle is given by the base times the height cubed divided by twelve. I_y is calculated in a similar fashion about the y-axis. Since the product of inertia of the rectangular areas A_1 and A_2 are zero relative to axes through their centroid parallel to xy then I_{xy} for the Z section is simply twice the area of A2 times the x and y distances from its centroid to the centroid of the z section. Carrying out these calculation results in I_x = 93.66 in^4, Iy = 17.66 in^4 and Ixy = 29.75 in^4 as shown in Figure 3-10.

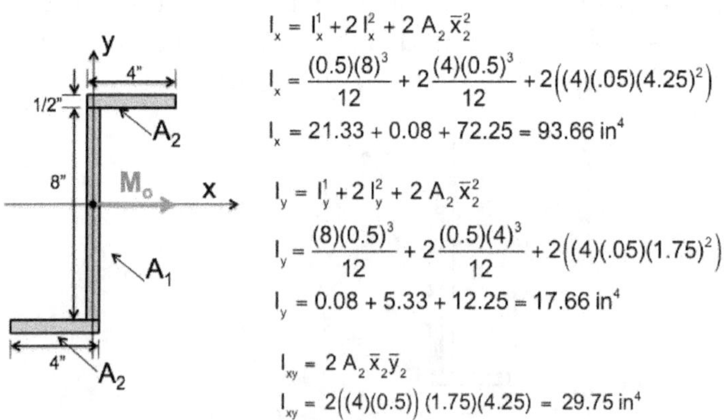

Figure 3-10 Calculation of I_x, I_y and I_{xy}

Calculate I_1, I_2 and Principal Axes

Mohr's circle is again used to calculate the principle moments of inertia of the Z section and the orientation of the principle axes. The numerical calculations for I_1 and I_2 are shown on the right side of Figure 3-11 using the generic formulas developed in Chapter 2. This results in $I_1 = 104\ in^4$ and $I_2 = 7.47\ in^4$. Applying the generic formula for determining the angle of rotation of the principle axis, θ, results in a counter clockwise rotation of 19 degrees.

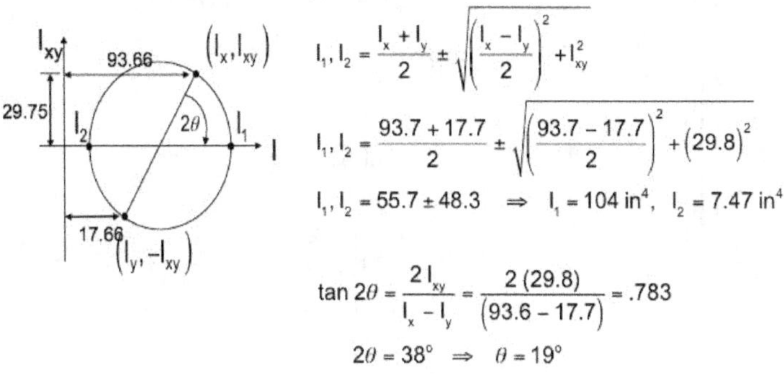

Figure 3-11 Calculation of I_1, I_2 and θ

Neutral axis and Max stress

Using the procedure developed in Chapter 2 for determining the angular position of the neutral axis relative to the principle axes results in an angle alpha, α of 78 degrees counter clockwise from the x' axis in Figure 3-12. It is also observed in Figure 3-12 that that the maximum stress will occur at the top left end of the upper leg of the Z section.

The x' and y' coordinates of this location are determined approximately from the sides of the gray right triangle whose hypotenuse is 4.5 inches. Note that the x' coordinate is negative making both contributions to the stress positive. The magnitude of the maximum stress with M_o in the units of inch pounds is 0.104 M_o psi, pounds per square inch.

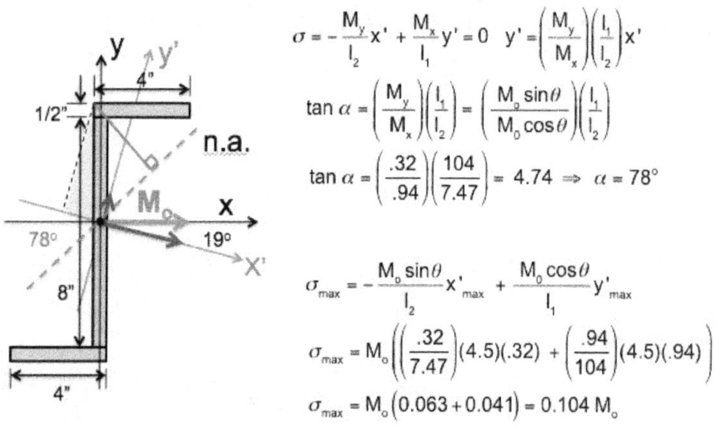

Figure 3-12 Neutral Axis and Maximum Stress

Comparison with Simple Beam Theory

Applying simple beam theory to approximate the maximum stress using My/I with y equal to 4.5 inches and I equal to 93.7 in^4 results in a value of 0.048M_o (see Figure 3-13). The percentage difference between the exact value of σ_{max} and the approximate value from My/I indicates that the approximate value is 53 percent lower than the exact value.

The difference in this example, where bending takes place about axes in a section that has no symmetry, is even greater than in Example 1.

$$\sigma = \frac{M}{I}y = M_o\frac{(4.5)}{(93.7)} = 0.048\ M_o$$

$$\frac{\sigma_{exact} - \sigma_{approx}}{\sigma_{exact}} = \frac{0.104 - 0.048}{0.104} = 0.53$$

$$\sigma_{approx} \cong 53\ \%\ \text{low}$$

Figure 3-13 Comparison with Simple Beam Theory

Design for Torsion

Chapter 1 Circular and Rectangular Cross Sections

Chapter 2 Hollow Non-Circular Cross Sections

Chapter 3 Practical Example Application

Chapter 1 – Circular and Rectangular Cross Sections

Chapter 1 of Design of Torsion covers a review of circular shaft torsion, the classic elasticity torsion problem, the membrane analogy solution for solid rectangular cross section shafts, comparison with the classic elasticity solution and an application to an H section extrusion.

Circular Shaft In Torsion

The deformation model assumed for the torsion of a solid or hollow circular shaft is that plane cross sections perpendicular to the z-axis remain plane when a torque T is applied about the central z-axis. This results in a shearing distortion of a cubic element oriented with its sides perpendicular to the xyz axis system in the xz plane and the generation of τ_{xz} shear stresses as depicted in Figure 1-1. This is characterized by the xy planes of the element remaining parallel to one other while the distortion takes place in the xz plane

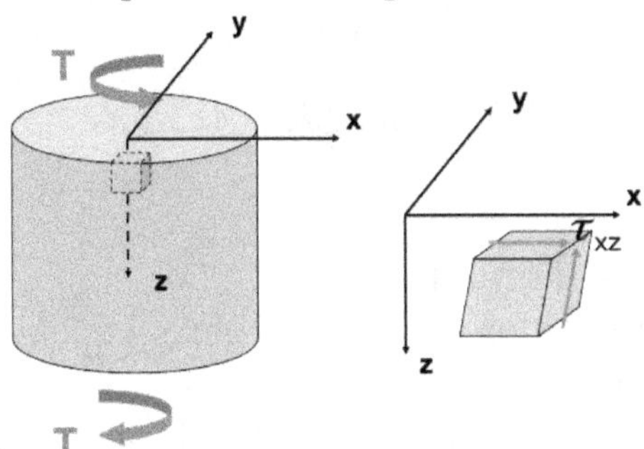

Figure 1-1: Circular Shaft In Torsion

Circular Shaft Analysis

The magnitude of the shear stress τ_{xz} is given by the equation τ equal to the applied torque, T, times the radial distance r measure from the center of the cross section divided by the polar moment inertia, J, of the circular cross section. The stress distribution varies linearly from zero at the center of the shaft to TR/J at the outside radius. The angular twist per unit length theta is given by the applied torque T divided by J times the shear modulus of the material G, i.e. T/JG. The total angular twist of a shaft φ of length L is the unit twist multiplied by the length.

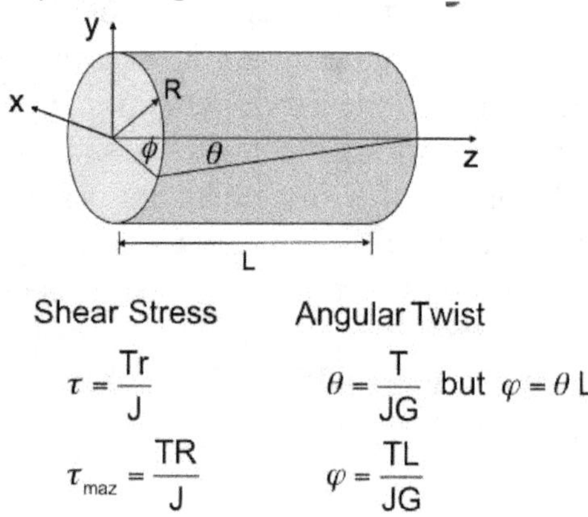

Figure 1-2: Circular Shaft Analysis

Square Shaft In Torsion

The deformation model used for a circular shaft is no longer applicable for a non-circular cross section shaft. In addition to distortion of the cubical element in the xz plane the xy planes also twists giving rise to a τ_{yz} stresses in addition to τ_{xz} stresses (see Figure 1-3). The result of this additional distortion is that an xy cross section that was originally plane now becomes warped. This complicates the torsion analysis beyond what can be solved with a simple model.

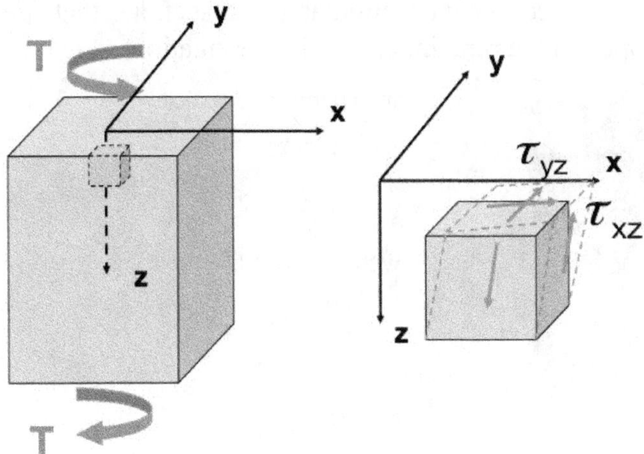

Figure 1-3: Square Shaft In Torsion

Twisted Square Shaft

Illustrated in Figure 1-4 is a photograph of a twisted square shaft on which a grid of mutually perpendicular lines was scribed before the shaft was twisted. It is clearly observed that cross sections that were original plane have now become warped. An exact solution of this problem makes use of the theory of elasticity. Its solution requires the integration of a second order partial differential equation whose result is a stress function, phi, φ. Once this function is obtained for a prescribed cross sectional geometry it derivatives with respect to x and y yield the two stress components τ_{yz} and τ_{xz}. The torque that produces the angle of twist, θ, is given by two times the integral of the stress function over the cross section.

This problem has been solved for a general rectangular cross section but for little else. This approach however does provide a basis for obtaining approximate solutions to certain non-circular cross sections that are developed in the remainder of this presentation.

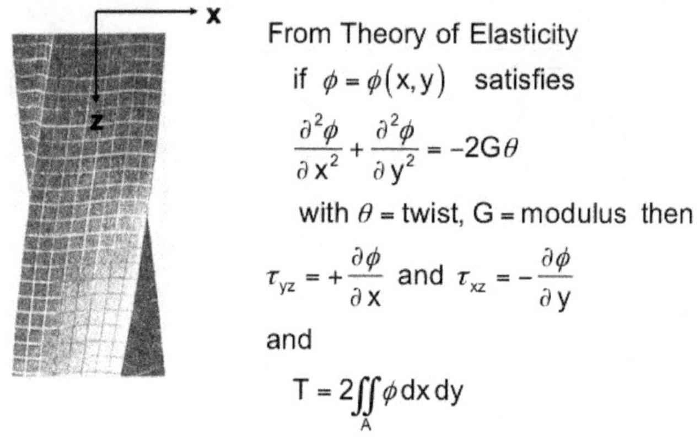

Figure 1-4: Twisted Square Shaft

Membrane Analogy

A problem which is mathematically similar to the torsion of an irregular non-circular shaft is that of a very flexible membrane, like a soap bubble, stretched across an opening on a flat plate whose shape is that of the cross section in question. If a small pressure difference is applied across the surface of the membrane it will undergo a vertical deflection that is governed by the same second order partial differential equation as that of the torsion problem. The only difference is that the right side of the membrane deflection equation is minus the pressure p times the tension, T_m, due to stretch in the membrane where as in the torsion problem the right side is -2Gθ. Illustrated by the flat gray plate in Figure 1-5

And the light gray dome is the deflected behavior of the pressurized membrane stretched over an elliptical hole in the plate.

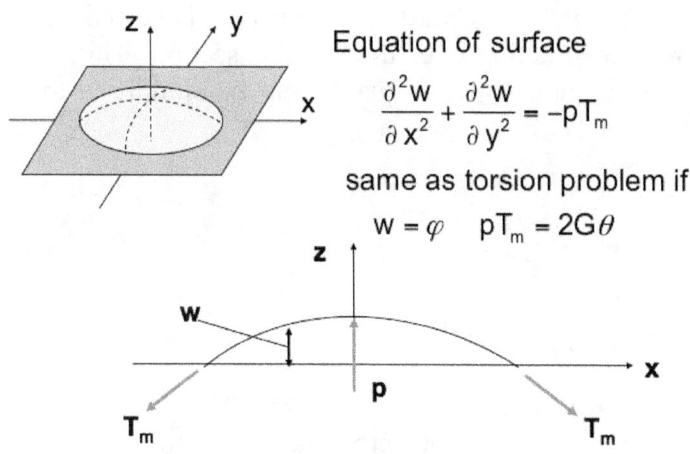

Figure 1-5: Membrane Analogy

A free body diagram of a section of the deflected membrane under the action of the tension T_m and the pressure p is shown at the bottom of Figure 1-5. This analogy will be used to solve the problem of the torsion of a circular shaft to demonstrate its applicability to analyze other cross sections of interest.

Membrane Analogy - Circular Shaft

To use the membrane analogy to solve the problem of the torsion of a solid circular shaft vertical equilibrium is applied to a section of the membrane highlighted in Figure 1-6. The net downward force on this section is the vertical component of the tension T_m around the circumference at radius r. This is expressed as $T_m(-dw/dr)2\pi r$. For small displacements the angle between the direction of T_m and the vertical can be approximated by -dw/dr.

This net downward force must be balanced by the pressure p acting over the circular section. This is expressed as $p(\pi r^2)$. This equality can be simplified to $dw/dr = -(p/2T_m)r$.

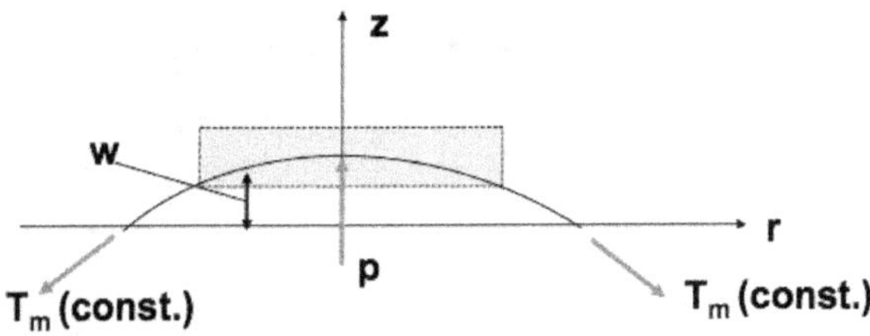

Vertical equilibrium (high lighted sec tion)

tension component = pressure component

$$(T_m)(-\frac{dw}{dr})2\pi r = p(\pi r^2) \Rightarrow \frac{dw}{dr} = -\frac{p}{2T_m}r$$

Figure 1-6: Membrane Analogy (Circular Shaft)

Integrating this first order equation in Figure 1-7 gives $w = -pr^2/4T_m$ plus a constant. At r = R, the radius of the shaft, the deflection of the membrane, w, is zero. The constant of integration becomes $pR^2/4T_m$. The final solution for w becomes Gθ/2 times $(R^2 - r^2)$ with 2Gθ substituted for p/T_m. This is done to make w the stress function φ for the theory of elasticity torsion problem. With τ_{xz} given as –dw/dr the maximum shear stress τ becomes Gθ/R. Theta, θ, now needs to be eliminated from this equation in terms of the applied torque T.

Integrating gives

$$w = -\frac{pr^2}{4T_m} + \text{const.}$$

$$\text{at } r = R \;\; w = 0 \;\; \Rightarrow \text{const.} = \frac{pR^2}{4T_m}$$

$$\text{and } w = \frac{p}{4T_m}\left(R^2 - r^2\right) = \frac{G\theta}{2}\left(R^2 - r^2\right)$$

$$\text{so that } \;\; \tau = -\frac{dw}{dr} = G\theta r \;\text{ and }\; \tau_{max} = G\theta R$$

Figure 1-7: Membrane Analogy (Circular Shaft)

Returning again to the elasticity solution for torsion the torque applied to the shaft can be represented as 2 times the integral of the stress function, φ, over the cross section. This integral is the same as the volume under the membrane that is given by the integral from 0 to R of w(2πrdr). Setting this equal to T/2 and carrying out the integration gives T = ($\pi R^4/2$) G θ (see Figure 1-8).

But J, the polar moment of inertia of a solid circular cross section is ($\pi R^4/2$) therefore θ = T/JG, the unit angle of twist derived from a strength of materials approach. τ_{max} then becomes T/RJ. The application of the membrane analogy gives the same result as the model that assumes that plane cross sections before torsion remain plane after twisting has occurred.

Volume under membrane

$$\text{Volume} = \int_0^R w(2\pi r)\,dr = \pi G\theta \int_0^R \left(R^2 r - r^3\right)dr = \frac{T}{2}$$

$$T = 2\pi G\theta\left[\frac{R^2 r^2}{2} - \frac{r^4}{4}\right] = \frac{\pi R^4}{2} G\theta$$

$$\theta = \frac{T}{JG} \quad \text{since} \quad J = \frac{\pi R^4}{2}$$

$$\text{and} \quad \tau_{max} = G\theta R = \frac{TR}{J}$$

Figure 1-8: Membrane Analogy (Circular Shaft)

Membrane Analogy – Thin Rectangular Section

The membrane analogy will now be used to obtain an approximate solution to the torsion of a thin rectangular cross section shaft about its longitudinal axis z as shown in Figure 1-9. The deflection of a membrane placed over a thin rectangular cross section hole in a flat plate where the dimension t is small compared to the dimension b can be approximated by a shape that will only be a function of x except very close to the ends where $y = \pm b/2$. This exception will only slightly affect the solution to the problem.

Equilibrium is again applied to a section of the deflected shape as illustrated in Figure 1-9. With T_m assumed constant and the ratio of b /t very large the net vertical down force per unit length in the z direction will be $-2T_m(dw/dx)$. This is balanced by the upward force from the pressure of p(2x). Simplifying this equality results in the first order differential equation that dw/dx = -(P/ T_m) x. Integrating from minus to plus t/2 results in w = Gθ ((t^2/4) – $x^{2)}$.

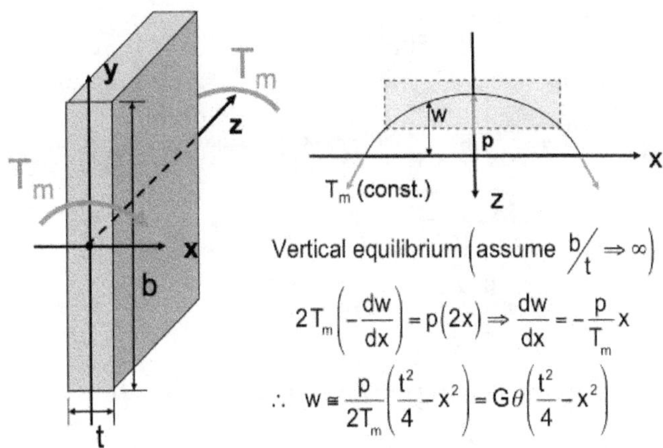

Figure 1-9: Membrane Analogy (Thin Rectangular Section)

Twist & Shear Stress

In Figure 1-10 the volume under the membrane that is equal to T/2 can be written as t^2 times the integral of wbdx from 0 to t/2. This results in the unit twist expression θ = 3T/ Gbt^3. The shear stress is again the derivative of w, the stress function, with respect to x which results in τ = 2Gθ x. Eliminating θ from the unit twist relationship gives τ = 6 Tx/ bt^3 with τ_{max} = 3T/ bt^2. Recall that these simple solutions for twist and stress are approximations that are only as good as the assumption that the shape of the membrane at the ends of the section where y = ± b/2 do not have a large effect.

Volume under membrane

$$V = 2\int_0^{t/2} wb\,dx = 2bG\theta\left[\frac{t^3}{8} - \frac{t^3}{24}\right] = bG\theta\frac{t^3}{6}$$

$$V = \frac{T}{2} \quad \Rightarrow \theta = \frac{3T}{G\left(bt^3\right)}$$

Shear Stress

$$\tau = -\frac{dw}{dx} = -\frac{d}{dx}\left[G\theta\left(\frac{t^2}{4} - x^2\right)\right] = 2G\theta x$$

$$\text{but } G\theta = \frac{3T}{\left(bt^3\right)}$$

$$\therefore \quad \tau = \frac{6Tx}{\left(bt^3\right)} \quad \Rightarrow \quad \tau_{max} = \frac{3T}{\left(bt^2\right)}$$

Figure 1-10: Twist & Shear Stress

Rectangular Section - Elasticity

The magnitudes of the multiplying constants C_1 and C_2 for the max stress and unit angle of twist as obtained from an exact theory of elasticity solution for a rectangular cross section are listed in Figure 1-11. It is observed that the membrane solution for this problem only corresponds to b/t equal to infinity for the elasticity solution.

If the ratio of b/t is 10 the simple membrane solution is in error about 6 percent. When compared to a square cross section with b/t equal to one the error is more like 38 percent for the stress and even greater for the twist.

$$\tau_{max} = c_1 \frac{T}{bt^2} \qquad \theta = c_2 \frac{T}{G(bt^3)}$$

b/t	infinite	10.00	5.00	3.00	2.00	1.00
c_1	3.00	3.20	3.44	3.74	4.06	4.80
c_2	3.00	3.21	3.44	3.80	4.37	7.09

Figure 1-11: Rectangular Section

Equivalent Sections

If the b/t ratio is large enough for the membrane solution to provide an acceptable result the shape of the arrangement of the cross section can be irrelevant to the

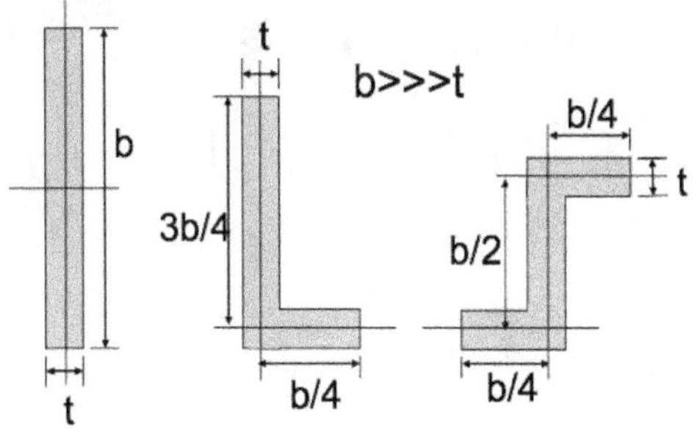

Figure 1-12: Equivalent Sections

maximum stress and unit angle of twist or torsional stiffness. As shown in Figure 1-12, whether the cross section is rectangular, L shaped or Z shaped the maximum stress and torsional stiffness in all three will be essentially the same if the total b dimension is the same for each cross section. The only correction required is for a stress concentration at an internal corner.

Stress Concentration

At the inside corners of the L and Z section the maximum stress is approximately equal to the actual stress times 1.74 $(t/r)^{1/3}$. In this correction r is the radius of the corner. In the example if r is 1/2 t the correction is 2.19 making τ_{max} a little greater than double that predicted by the membrane analogy solution for a rectangular cross section.

At inside corners – L & Z sections

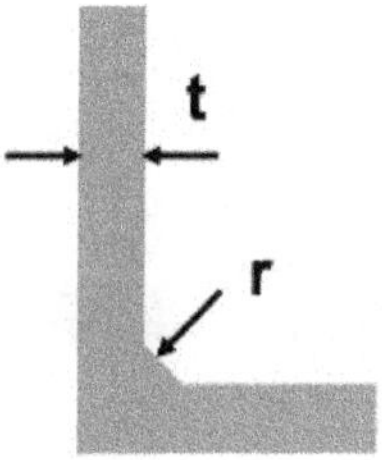

t

r

$$\tau_{max} = \tau_{rect} 1.74 \left(\frac{t}{r}\right)^{1/3}$$

Example: $r = t/2$

$$\tau_{max} = \tau_{rect} 1.74 \left(\frac{t}{r}\right)^{1/3} = \tau_{rect} 1.74 \sqrt[3]{2}$$

$$\tau_{max} = \tau_{rect} 1.74 \times 1.26 = 2.19$$

Figure 1-13: Stress Concentration

Sample Problem

The membrane analogy solution for a thin rectangular shaft will now be applied to a problem involving a much more complex cross section. This example deals with an Aluminum extrusion H cross section with the dimensions shown in Figure 1-14.

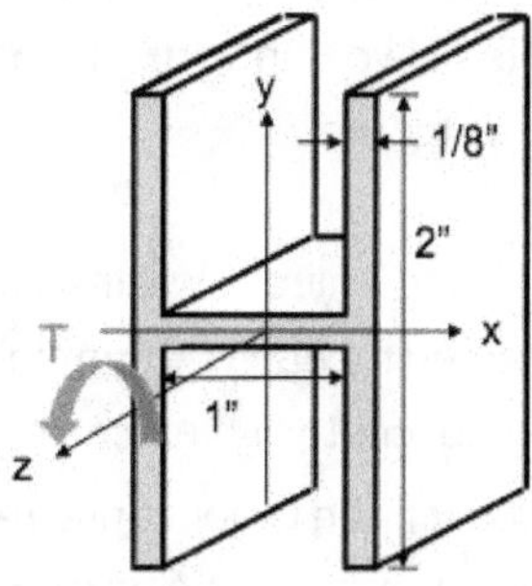

Figure 1-14: Sample Problem

The thickness of all webs is 1/8". It is desired to determine the magnitude of the torque that can be applied about the z-axis for a maximum allowable shear stress of 15,000 psi. Also determine the unit angle of twist. Stress concentration effects need not be included.

To use the membrane analogy method of solution it is necessary to visualize what the shape of a membrane would look like if pressurized over a cut out of the form of the H cross section. Since the web is only 1/8 inch thick compared to the vertical and cross web element dimensions the shape would be very similar to that assumed for a single rectangular cross section except in the immediate vicinity of the intersection of the cross web with the vertical elements. Based on this observation it is a reasonable assumption that the torque carried by the H section can be treated as a composite of two thin vertical rectangles and a thin horizontal rectangle.

Based on membrane analogy H section can be treated as a composite of two thin vertical rectangles and a thin horizontal rectangle.

The torque carried by the entire H will then be the sum of the torques carried by each rectangle.

$T = 2\,T_v + T_h$ (subscript v is vertical rectangle, subscript h is horizontal rectangle)

and all three rectangles will rotate the same amount

$\theta = \theta_v = \theta_h$

Figure 1-15: Solution

This is expressed in Figure 1-15 as T equal to $2T_v + T_h$. All three rectangles will of necessity have to rotate the same amount. Hence, θ for the H section is equal to θ_v equal to θ_h.

Shear & Torque

From the membrane analogy torsion solution for a narrow rectangular cross section the shear stress is given by three times the torque applied to that rectangle divided by b the long dimension of the rectangle times t squared (see Figure 1-16). "t" is the narrow dimension of the rectangle. Solving this equation for the torque gives $T = \tau bt^2/3$.

The stress for both the vertical and cross elements of the section will be the same as will the thickness t. Taking this into account in adding the torque components of the individual rectangles for the total torque results in the equation $T = (\tau_{max}t^2/3)\,(2b_v + b_h)$. Substituting the appropriate values of the individual parameters into the equation for the total torque results in a magnitude of 390 in. lbs.

$$\tau = \frac{3T}{bt^2} \Rightarrow T = \frac{\tau b t^2}{3}$$

For H section

$$T = 2T_v + T_h$$

$$T = \frac{\tau_{max} t^2}{3}(2b_v + b_h)$$

$$T = \frac{(15x10^3)(.125)^2}{3}(4+1)$$

$$T = 390 \text{ in. lb.}$$

Figure 1-16: Shear & Torque

Unit Angle of Twist

From the membrane analogy torsion solution the unit angle of twist is given by $\theta = 3T/Gt_3$. By eliminating the term T/bt^2 between the equation for the angle of twist and the shear stress a simplified equation is obtained for theta equal to $\theta = \tau/Gt$ in Figure 1-17. Substituting the appropriate parameter values into this equation gives a final result of .032 radians or 1.83 degrees. It is of interest to note that only the maximum shear stress, the material property and the narrow dimension of the rectangle are need to determine the angular twist. With τ_{max} and t the same for all rectangular elements in the H section this same equation is applicable to all three elements.

$$\tau = \frac{3T}{bt^2} \qquad \theta = \frac{3T}{G\left(bt^3\right)} \text{ unit twist angle}$$

Eliminate $\frac{T}{bt^2}$ btween stress and twist angle

$$\theta = \frac{\tau}{Gt} = \frac{\tau_{max}}{Gt}$$

with shear modulus $G = 3.8 \times 10^6 \text{ lb/in}^2$

$$\theta = \frac{15 \times 10^3}{\left(3.8 \times 10^6\right)\left(.125\right)} = .032 \text{ rad} = 1.83 \text{ deg}$$

Figure 1-17: Unit Angle of Twist

Chapter 2 – Hollow Non-Circular Cross Sections

Introduction

Chapter 2 of Design for Torsion covers the application of the membrane analogy for torsion to hollow thin wall non-circular cross sections shafts with an introduction to the analysis of multi-cell tubes.

Shear Flow Introduction

The concept of shear flow, q, is introduced in Figure 2-1 for a thin wall hollow non-circular cross section shaft. It is defined as the product of the shear stress, τ, and the thickness of the wall of the hollow section, t. The units of shear flow are lbs. per inch and q is a constant around the perimeter of the section.

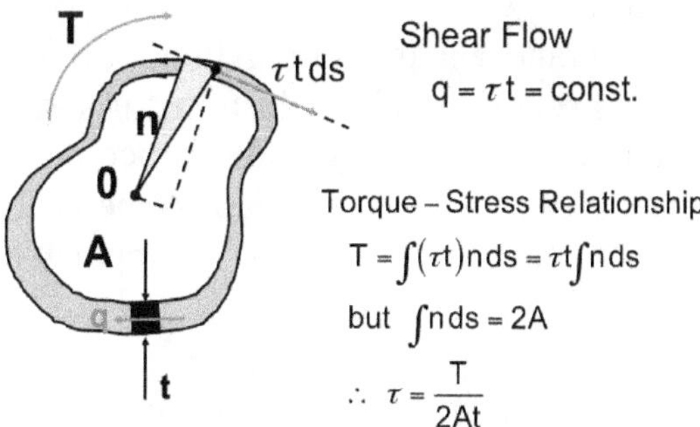

Figure 2-1 Hollow Non-Circular Section

The applied torque T can be related to the stress by recognizing that over increment ds of the perimeter the increment of force generated by the shear stress can be expressed as qds or τtds.

If this incremental force, multiplied by the perpendicular distance, n, from the line of action of the force to some point O in the section, is integrated around the perimeter this will be equal to the applied torque, T. This integral is represented as qnds or τtnds. Since the shear flow q is constant it can be taken outside the integral leaving nds around the perimeter (see Figure 2-1). This integral can be interpreted as twice the area of the light gray triangle summed around the perimeter so its value is just twice the enclosed area of the cross section A.

Combining this with the equation for the torque results in the shear stress, τ, given as T/2At, where A is the enclosed area and t is the thickness of the wall of the hollow section.

Membrane Analogy

The membrane analogy for a thin wall hollow non-circular cross section shaft is illustrated in Figure 2-2. A hole is cut in the light gray plate that conforms to the outside perimeter of the hollow shaft. A membrane is placed over this opening as in the earlier application in Chapter 1.

A second plate, shown in darker gray, somewhat smaller than the hole cut in the base plate is placed on top of the membrane leaving only the narrow remaining portion of the membrane representing the wall of the hollow section exposed to be deformed upward by the differential pressure across the membrane. An analysis of the deformation of this part of the membrane will lead to a relationship of the applied torque to the unit angle of twist.

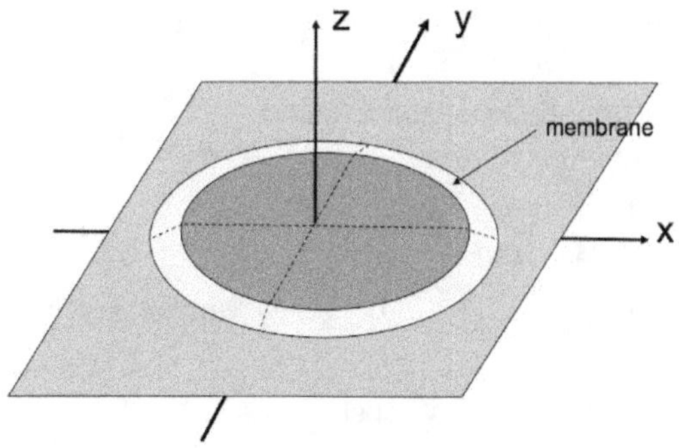

Figure 2-2 Membrane Analogy

Membrane Analysis – Hollow Shaft

As in the previous analysis of the membrane model behavior vertical equilibrium will be applied to the distorted membrane shape representing the non-circular cross section shaft. If the wall thickness is small compared to the equivalent diameter of the shaft the upward force can be represented by the pressure p times the enclosed area of the section, A. This will be balanced by the downward component of force from the tension in the membrane as shown in Figure 2-3. This can be represented by the integral of T_m (h/t) ds.

Setting $p/T_m = 2G\theta$ and $\tau = h/t$, the slope of the stress function, the equation of vertical equilibrium becomes $2G\theta = 1/A$ times the integral of τds. But τ is also given by T/2At. Solving for θ gives the constant $T/4GA^2$ times the integral of ds/t. assuming that t is constant then $\theta = TL/4GA^2t$. L in this equation is the length around the perimeter of the cross section. For noncircular cross sections in torsion with a constant wall thickness expressions for the unit twist and the stress have now been determined.

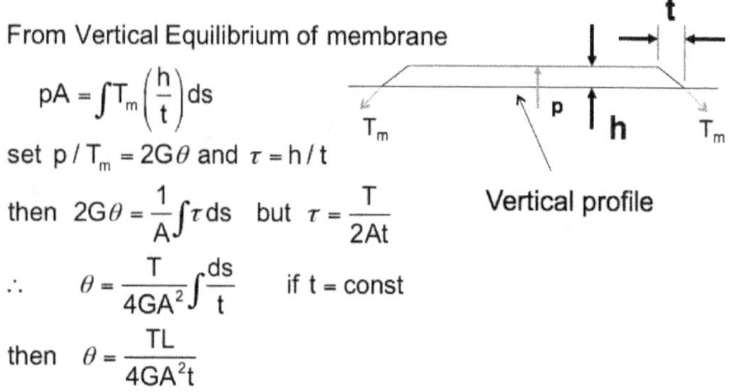

Figure 2-3 Torsion of Non-Circular Section

Thin Wall Circular Shaft

The shear stress for a circular cross section is given by $\tau = Tr/J$. If t is small compared to R then r is approximately R and J can be represented as $2\pi Rt(R^2)$, see Figure 2-4. Substituting this into the stress equation gives $\tau = T/2\pi R^2 t$.

Now solve for the stress in this section using the result obtained from the membrane analogy solution where $\tau = T/2At$. With A equal to πR^2 the shear stress is $\tau = T/ 2\pi R^2 t$. The same result is obtained using the two different solution methods. This provides credibility for the more general membrane solution formulation.

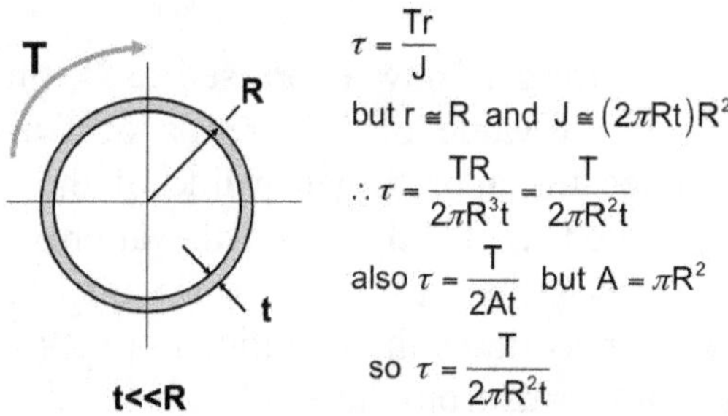

Figure 2-4 Thin Wall Circular Shaft

Exact Thin Wall Circular Shaft

An interesting exercise is to determine the magnitude of the error in the approximate stress calculated in Figure 2-4 with the exact solution for the stress. The exact stress magnitude is calculated first.

Beginning with τ =Tr/J little r is set equal R. J is expressed as π/2 times R^4 minus the quantity $(R-t)^4$. This is the exact value of J (Figure 2-5).

$$\tau = \frac{Tr}{J}$$

$$\text{where } r = R \text{ \& } J = \pi/2\left[R^4 - (R-t)^4\right]$$

$$J = \pi/2\left[R^4 - \left(R^4 - 4R^3t + 6R^2t^2 - 4Rt^3 + t^4\right)\right]$$

$$\text{or } J = \pi/2\left(4R^3t\right)\left(1 - 3/2\left(\frac{t}{R}\right) + \left(\frac{t}{R}\right)^2 - 1/4\left(\frac{t}{R}\right)^3\right)$$

Figure 2-5 Exact J Value

This can be simplified to J equal to π/2 times $4R^3t$ times the quantity $1 - (3/2)(t/R) + (t/R)^2 - ¼\ (t/R)^3$.

The exact stress is now expressed as T time R divided by the exact value of J in terms of t and R (Figure 2-6). Since the approximate value of the shear stress was calculated as $T/2\pi R^2 t$ then the approximate shear can also be written as the exact value of the maximum shear stress times the quantity 1- (3/2)(t/R) + the additional (t/R) terms from the exact value of J.

$$\tau = \frac{TR}{\pi/2\left(4R^3 t\right)\left(1-3/2\left(\frac{t}{R}\right)+\left(\frac{t}{R}\right)^2-1/4\left(\frac{t}{R}\right)^3\right)}$$

$$\tau = \frac{T}{2\pi R^2 t\left(1-3/2\left(\frac{t}{R}\right)+\left(\frac{t}{R}\right)^2-1/4\left(\frac{t}{R}\right)^3\right)}$$

$$\tau_{app} = \tau_{exact}\left(1-3/2\left(\frac{t}{R}\right)+\left(\frac{t}{R}\right)^2-1/4\left(\frac{t}{R}\right)^3\right)$$

Figure 2-6 Exact & Approximate Stress

Numerical Comparison

The table in Figure 2-7 lists the magnitude of the approximate shear stress over the exact shear stress for increasing values of the ratio of t/R. It is observed that for t/R = 0.1 or t is 10% of R the approximate solution is 14 percent below the actual stress. For lower values of t /R the error is less. The value of shear stress from the approximate formulation is always less than the exact value. It is probably not too unreasonable to use the approximate formulation for value of t up to 10 percent of R particularly in those geometries where no more accurate solution is available.

$$\tau_{app} = \tau_{exact}\left(1 - \frac{3}{2}\left(\frac{t}{R}\right) + \left(\frac{t}{R}\right)^2 - \frac{1}{4}\left(\frac{t}{R}\right)^3\right)$$

t/R	App/Exact	% of R
0.02	0.97	2
0.04	0.94	4
0.06	0.91	6
0.08	0.89	8
0.1	0.86	10
0.12	0.83	12
0.14	0.81	14

Figure 2-7 Numerical Comparison

Comparative Cross-Sections

The effect of changes in geometry of non-circular sections can have a dramatic effect on their comparative torsional characteristics. Consider the torsional stiffness of a thin wall hollow tube as measured by the ratio of the applied torque, T, to the unit angle of twist, θ. From membrane theory this is given by the expression $4GA^2t/L$. This will be used as a basis of comparison beginning with the circular cross section shown in the upper left of Figure 2-8. If this tube is flattened into an elliptical cross section while keeping L and t constant the only property that changes is the included area A which is less than for a circular cross section. This in turn means the torsional stiffness will also be lower than that of the circular cross section. The result is that the shaft will twist to a greater angle θ for the same applied torque T even though the same amount of material is used in the shaft. If this elliptical cross section shaft is flattened even more with t and L remaining constant the torsional stiffness will be reduced even further as A for the flattened geometry is now significantly less than for the round and elliptical cross section geometries.

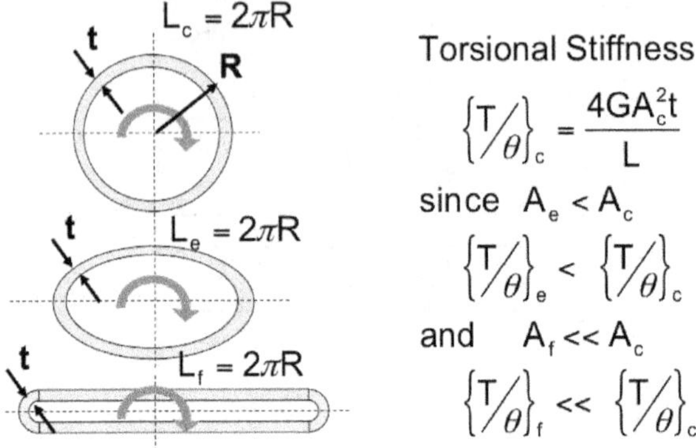

Figure 2-8 Comparative Cross-Sections

Square and Round Tubes

Another interesting comparison is the difference in the maximum shear stress generated in a square hollow tube and a round hollow tube each containing the same amount of material, see Figure 2-9. To satisfy this condition the volume of material in the square hollow tube which is 4Lt must be equal to 2πRt for the circular hollow tube. This requires L to be πR/2. The shear stress in the circular tube is given by $T/2\pi R^2 t$. For the square tube the shear stress is given by $T/2A^2 t$. Substituting the value of L required for equal material usage in the square tube the shear stress becomes $T/2\pi R^2 t$ times the quantity (4/π). This results in a shear stress in the square tube almost 30 % higher than the stress in the round tube for the same applied load not taking into account any stress concentration effect. It is left for the viewer to show how the torsional stiffness of the square tube differs from that of the circular tube.

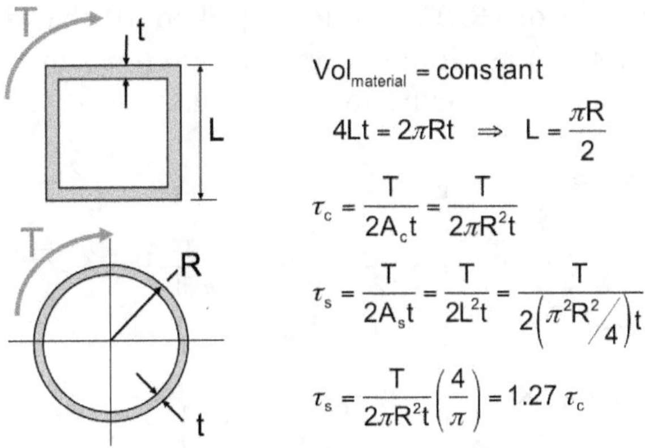

Figure 2-9 Square & Round Tubes

Slit Tube Comparison

Consider how the stress levels and torsional stiffness compare for a thin wall circular tube slit along it entire length as depicted in Figure 2-10 to a completely closed thin wall circular tube subjected to the same external torque, T. The stress in the slit tube is calculated from the membrane analogy solution for a thin rectangular cross section. With b equal to about $2\pi R$ the shear stress τ_s becomes $(3/2)\ T/\pi Rt^2$. For the closed circular section the stress τ_c is T/2At.

With A equal to πR^2 the shear stress τ_c becomes ½ (t/R) times the quantity $(T/\pi\ Rt^2)$ (see Figure 2-10). A comparison of these two shear stress values gives $\tau_s = 3(R/t)\ \tau_c$, the shear stress in the closed circular tube. If the thickness t is one tenth of the radius R the shear stress in the slit tube would be 30 times higher than in the closed tube for the same applied torque. The unit angle of twist for the slit tube can be shown to be equal to three times the unit angle of twist for the closed tube times the ratio of $(R/t)^2$. For R/t equal to 10 the unit angle of twist in the slit tube would be 300 times that of the closed tube for the same applied torque.

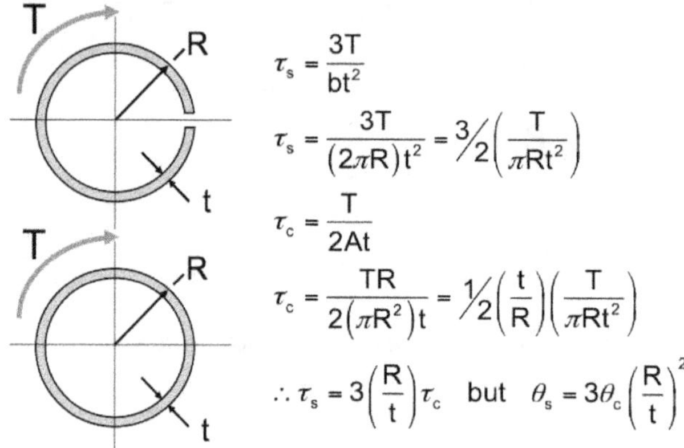

Figure 2-10 Slit Tube Comparison

These dramatic differences are a consequence of the slit tube really behaving like a very thin rectangular cross section shaft having been bent into a circular shape.

Membrane Solution Summary

Listed in Figures 2-11 and 2-12 are the shear stress and unit angle of twist formulas for a thin rectangular cross section shaft and a thin wall hollow non-circular cross section tube derived from the application of the membrane analogy for torsion.

For a thin rectangular cross section shaft

Shear stress $\tau_{max} = \frac{3T}{(bt^2)}$

Unit twist angle $\theta = \frac{3T}{G(bt^3)}$

where t = width of rectangle, b = height of rectangle

Figure 2-11 Thin Rectangular Section

These are approximate solutions that depend on certain restrictive assumptions. For the thin rectangular cross section shaft the ratio b/t needs to be large.

For a thin walled non-circular cross section tube

Shear stress $\tau = \frac{T}{2AL}$

Unit twist angle $\theta = \frac{TL}{4GA^2t}$

where t = wall thickness, A = enclosed area, L = length of perimeter

Figure 2-12 Thin Non-Circular Section

For the thin wall hollow non-circular cross section the wall thickness t is assumed to be small compared to some equivalent tube radius and is taken to be constant around the perimeter of the cross section. Even with these restrictions these formulae can be very useful for a variety of geometries that are non-circular.

Two Cell Systems

The membrane analogy results can be applied with some additional restrictions to the solution of the torsional behavior of multiple cell thin wall tubes. The two-cell tube illustrated in Figure 2-13 possesses different enclosed areas A_1 and A_2 as well as three different wall thick-nesses t_1, t_2 and t_3. The problem is to determine the three shear stresses that will be generated in these three thicknesses by an applied torque T.

Shear flow, q, defined as τt, can be treated like an incompressible fluid in pipe flow. At the point C where three wall thicknesses converge this application of the shear flow concept gives rise to the equation that q_1-q_3 = q_2. This leads directly to equation (1) in Figure 2-13 relating the three shear stresses and the three thicknesses.

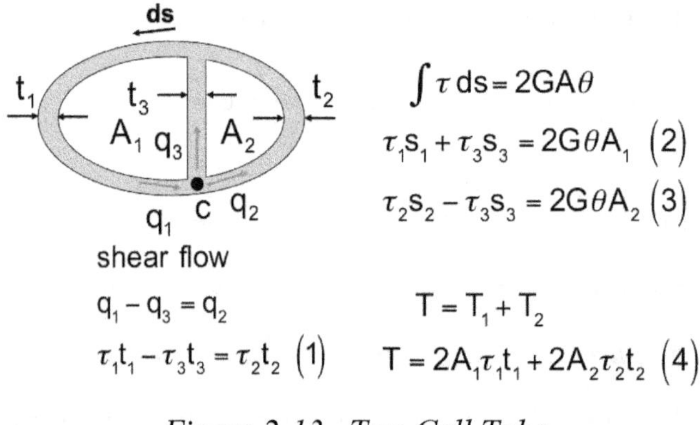

Figure 2-13 Two Cell Tube

The generic membrane analogy equation that states the integral of τds around the perimeter is equal to 2GθA gives rise to equations (2) and (3) where the value of s corresponds to the length of the perimeter associated with a particular wall thickness. Finally the total applied torque is the sum of the component torques carried by each cell. Representing these torque components in terms of their individual cell parameters results in equation (4). This system of equations is sufficient to determine the three unknown shear stresses and the unit angle of twist in terms of the applied torque.

Shear Stress Solution

Listed in Figure 2-14 are the equations defining τ_1, τ_2 and τ_3 in terms of the applied torque T as obtained from the simultaneous solution of the equation set in Figure 2-13.

$$\tau_1 = \frac{T\left\{t_3 s_2 A_1 + t_2 s_3 \left(A_1 + A_2\right)\right\}}{2\left\{t_1 t_3 s_2 A_1^2 + t_2 t_3 s_1 A_2^2 + t_1 t_2 s_3 \left(A_1 + A_2\right)^2\right\}}$$

$$\tau_2 = \frac{T\left\{t_3 s_1 A_2 + t_1 s_3 \left(A_1 + A_2\right)\right\}}{2\left\{t_1 t_3 s_2 A_1^2 + t_2 t_3 s_1 A_2^2 + t_1 t_2 s_3 \left(A_1 + A_2\right)^2\right\}}$$

$$\tau_3 = \frac{T\left\{t_1 s_2 A_1 - t_2 s_1 A_2\right\}}{2\left\{t_1 t_3 s_2 A_1^2 + t_2 t_3 s_1 A_2^2 + t_1 t_2 s_3 \left(A_1 + A_2\right)^2\right\}}$$

Figure 2-14 Shear Stress Solution

Note that each equation contains both enclosed areas A_1 and A_2, the three wall thicknesses t_1, t_2 and t_3 as well as the perimeter length s_1, s_2 and s_3 associated with the three corresponding wall thickness sections.

Special Case Example

This generic two-cell solution is now applied to the illustrated circular thin wall tube with a web across the diameter in Figure 2-15. Assuming that t_1 is equal to t_2, s_1 is equal to s_2, and A_1 is equal to A_2, the shear stress τ_3 in the web from Figure 2-14 is calculated to be zero.

Now set all thicknesses to t, s_1 and s_2 to πR, s_3 to 2 R and A_1 and A_2 to $\pi R^2/2$. Substituting these values into either the solution for τ_1 or τ_2 results in $\tau = T/2R^2t$. This is the value of the shear stress calculated earlier for a single circular thin wall tube. This means that the web across the diameter contributes nothing to the torsional characteristics of this special two-cell tube.

Symetrical Cross Section

$t_1 = t_2$, $s_1 = s_2$, $A_1 = A_2$

then $\tau_3 = 0$

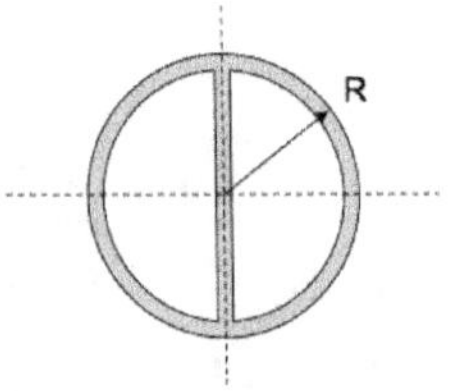

Circular Section with Web

$t_1 = t_2 = t$, $s_1 = s_2 = \pi R$, $s_3 = 2R$, $A_1 = A_2 = \pi R^2/2$

$\tau = \dfrac{T}{2\pi R^2 t}$ shear stress in circular thin walled tube

R

Center web has no effect

Figure 2-15 Special Case

Chapter 3 – Practical Application

Introduction

Chapter 3 of Design for Torsion deals with the determination of the torsional behavior of a proposed connecting tube design using non-circular hollow shaft theory.

Problem Statement

As illustrated in Figure 3-1 two storage tanks are to be joined together with a 10 in. long gastight coupling that provides for some torsional flexibility about the longitudinal axis of the coupling tube. It is proposed to use a thin walled connection with many convolutions as shown in Figure 3-2.

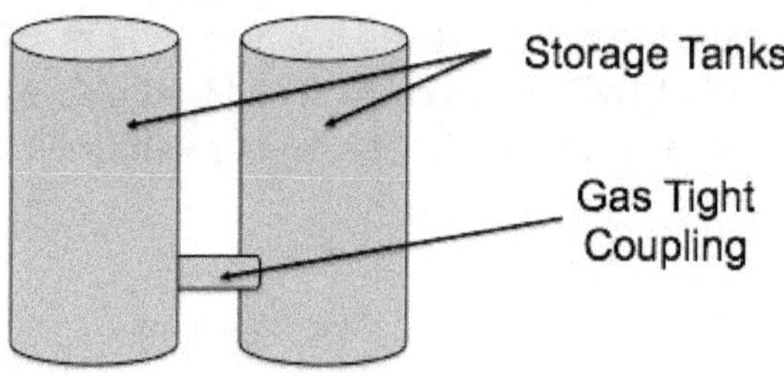

Figure 3-1 Storage Tanks

The connection is formed from bronze sheet with 30 fingers around the circumference. If the working stress is not to exceed 10,000 psi what is the permissible angle of twist. Compare this allowable twist with that of a circular tube of 5 in. diameter made of the same material.

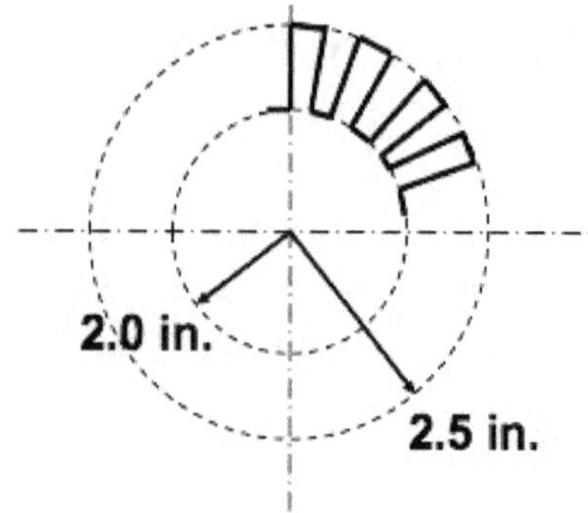

Figure 3-2 Connection Cross Section

Applicable Relationships

To solve this problem relationships for shear stress and angular twist as a function of applied torque for a thin wall noncircular tube are required. From the application of membrane analogy for torsion in Chapter 2 the formula for the shear stress is given by $\tau = T/2At$. The unit angle of twist is given by $\theta = TL/4GA^2t$ where L is the perimeter length L of the convolutions (see Figure 3-3).

Thin Walled Non – Circular Cross Section

Torsional Shear stress

$$\tau = \frac{T}{2At}, \text{ A = enclosed area, t = thickness}$$

Twist per unit length

$$\theta = \frac{TL}{4GA^2t}, \text{ L = length of convolutions}$$

Figure 3-3 Thin Walled Non-Circular Section

Relation of Shear to Twist

To relate the stress to the twist behavior the quantity T/t can be eliminated between the two equations in Figure 3-3. Carrying out this manipulation results in $\theta = L/2AG$ as shown in Figure 3-4. The enclosed area A and the perimeter length L must to be determined from the geometry of the convoluted tube.

Eliminate T/t from Stress andTwist Equations

from stress $$\frac{T}{t} = 2A\tau$$

so that $$\theta = \frac{L}{4GA^2}\left(\frac{T}{t}\right) = \frac{L}{4GA^2}(2A\tau)$$

$$\therefore \quad \theta = \tau\left(\frac{L}{2AG}\right)$$

Figure 3-4 Shear and Twist Relation

Determination of A & L

The enclosed area will simply be the area of a circle of r_i, the inner radius, plus one half the area of an annulus from r_i to r_o, the outer radius. These two terms combined result in $A = (\pi/2)(r_i^2 + r_o^2)$.

Enclosed Area A

$$A = \pi r_i^2 + \frac{1}{2}\left(\pi r_o^2 - \pi r_i^2\right) = \frac{\pi}{2}\left(r_o^2 + r_i^2\right)$$

r_i = inside radius, r_o = outside radius

Length of Convolutions L (N = number)

$$L = 2N(r_o - r_i) + \frac{1}{2}(2\pi r_o + 2\pi r_i)$$

$$L = 2N(r_o - r_i) + \pi(r_o + r_i)$$

Figure 3-5 A and L Determination

The perimeter length L of the convolutions can be expressed physically as 2N $(r_o - r_i)$ + ½ $(2\pi\, r_o + 2\pi\, r_i)$, as shown in Figure 3-5.

Calculation of A & L

The numerical values of the enclosed are A and perimeter length L are calculated by substituting the appropriate geometric dimensions into the equation developed in Figure 3-5. This results in A = 16.09 in.^{2} and L = 44.13 inches in Figure 3-6.

Enclosed Area A

$$A = \frac{\pi}{2}\left(r_o^2 + r_i^2\right) = \frac{3.14}{2}\left(2.5^2 + 2.0^2\right)$$

$$A = 16.09 \text{ in.}^2$$

$$L = 2N\left(r_o - r_i\right) + \pi\left(r_o + r_i\right)$$

$$L = 2x30\left(2.5 - 2.0\right) + 3.14\left(2.5 + 2.0\right)$$

$$L = 30 + 14.13 = 44.13 \text{ in.}$$

Figure 3-6 Calculation of A & L

Twist Calculation

The unit angle of twist is calculated as 0.00228 radians per inch of length in Figure 3-7. Multiplying this value by the length of the tube and converting radians to degrees gives a final value of 1.31 degrees.

This does not appear to be a very large. Perhaps it would be just as well to use a 5 inch diameter circular hollow tube and not bother with including all the convolutions.

Twist per unit length

$$\theta = \tau\left(\frac{L}{2AG}\right) = 10x10^3\left(\frac{44.13}{2x16.09x6x10^6}\right)$$

$$\theta = 2.28x10^{-3} = .00228 \text{ rad/in.}$$

Total Twist

$$\theta_t = \theta \, x \, \text{tube length} = .00228x57.3x10$$

$$\theta_t = 1.31 \text{ deg.}$$

Figure 3-7 Calculated Twist

Circular Section Replacement

To establish a comparison of the convoluted tube to a five inch circular cross section tube the equations for the shear stress and unit angular twist are listed in Figure 3-8 assuming that the wall thickness is small to the radius. With a reasonable wall thickness of say .050 inches the ratio of t to r would be 1 to 100 which justifies the thin wall approximation.

Thin Walled Circular Cross Section

Torsional Shear stress

$$\tau = \frac{T}{2\pi r_o^2 t}$$

Twist per unit length

$$\theta = \frac{T}{2\pi r_o^3 G t}$$

Figure 3-8 Circular Section

Twist In Terms of Stress

By again eliminating the ratio of T/t between the two equations in Figure 3-8 results in $\theta = \tau / r_o G$. Substituting the values of the parameters into the right side of the equation gives a unit twist of 0.67 x 10^{-3} radians per inch in Figure 3-9. The total twist in for the five inch circular tube is 0.38 degrees.

Eliminate T/t between Stress and Twist

from stress $\quad T/t = 2\pi r_o^2 \tau$

so that $\quad \theta = \dfrac{1}{2\pi r_o^3 G}\left(T/t\right) = \dfrac{1}{2\pi r_o^3 G}\left(2\pi r_o^2 \tau\right)$

$$\therefore \theta = \left(\frac{\tau}{r_o G}\right) = \frac{10 \times 10^3}{2.5\left(6 \times 10^6\right)} = 0.67 \times 10^{-3} \text{ rad/in.}$$

$$\theta_t = \theta \times \text{length} = .00067 \times 57.3 \times 10 = 0.38 \text{ deg}$$

Figure 3-9 Twist In Terms Of Stress

Comparison

The ratio of the total angle of twist of the convoluted tube to the circular tube is 1.31 divided by 0.38 or 3.45 in Figure 3-10. The convoluted tube is almost three and a half times more flexible in torsion than the circular cross section tube with the same wall thickness.

It is of interest to note that the allowable torque that can be carried by either of these tubes cannot be determined without first specifying their wall thicknesses. Also, even if the wall thickness were different from one another the ratio defining their relative torsional flexibility would not change.

$$\frac{\theta_t(\text{convoluted tube})}{\theta_t(\text{circular tube})} = \frac{1.31}{0.38}$$

$$\frac{\theta_t(\text{convoluted tube})}{\theta_t(\text{circular tube})} = 3.45$$

Almost $3\frac{1}{2}$ times more flexible

Figure 3-10 Comparison

Tank Movement

Assuming that the top of the tank is six feet above the coupling the distance one tank top can move relative to the adjacent tank by twist of the coupling can be determined as the product of the unit twist times the length of the coupling times the height of the tank all in the proper units. Carrying out his calculation in Figure 3-11 gives a maximum possible movement of two inches. This is the permissible movement in one direction. Since the coupling can also twist in the opposite direction the total permissible twist is 4 inches.

Assume top of tank is 6 ft. above coupling

Distance tank top can move due to the rotation of coupling

$$\delta = \theta \, L \, h$$

$$= .00228 \text{rad/in.} \times 10 \text{ inches} \times 6 \text{ ft.} \times 12 \text{ in./ft}$$

$$\delta = 2 \text{ in.}$$

Figure 3-11 Tank Movement

Design for Buckling

Chapter 1 Buckling Fundamentals

Chapter 2 Effect of Support on Euler Buckling Load

Chapter 3 Material Properties & Cross Sectional Geometry Effects

Chapter 4 Shortening of Columns

Chapter 5 Thermostat Redesign Problem

Chapter 1 – Buckling Fundamentals

Introduction

This portion of Mechanical Design Engineering Monograph V deals with the generic problem of how an axially applied load can produce lateral instability. The issues addressed include how the magnitude of this load is related to the support of the member, how material properties and cross sectional geometry govern instability, how lateral movement is related to the shortening of the member and an application of these materials in a real redesign problem.

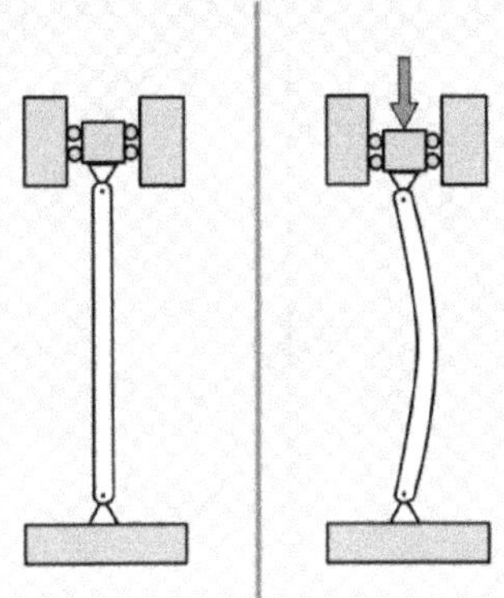

Figure 1-1 Slender Member

Column Behavior

When the length of an axially loaded compression member is large in comparison to its transverse dimensions it's bending behavior and potential failure can occur due to a buckling instability exhibited by sudden and excessive lateral displacement as illustrated in Figure 1-1. This deflection behavior and the limited load carrying capacity of the member can be initiated at stress levels below that of the compressive yield strength of the material. These members are called columns if they stand vertically or struts and braces if placed diagonally.

Axially Loaded Model

To develop an understanding of this unusual phenomena consider the lateral displacement of a column that is built in at its base and free to move laterally at it top. The applied axial load P acts at a small eccentricity, ε, from the centroid of the cross section of the member. This provides greater generality to the bending analysis.

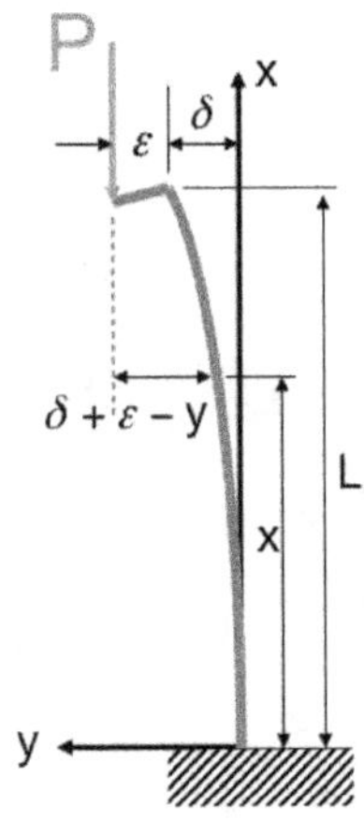

Figure 1-2 Axially Loaded Member

The bending moment in the member at the distance x measured from the base is expressed as $-P(\delta + \varepsilon - y)$. δ is the lateral displacement of the member at $x = L$, ε is the eccentricity of the load from the centroid of the cross section and y is the lateral displacement of the member at x. This expression for the bending moment is substituted into the differential equation EI $(d^2y/dx^2) = -M(x)$ that governs the lateral deflection of a member subjected to a bending moment distribution along its length as shown in Figure 1-3. E in this equation is the modulus of elasticity of the member and I is the area moment of inertia of its cross section.

Bending Moment

$$M(x) = -P(\delta + \varepsilon - y)$$

Deflection Differential Equation

$$EI\frac{d^2y}{dx^2} = -M(x) = P(\delta + \varepsilon - y)$$

Define

$$P/_{EI} = k^2 \qquad \text{then}$$

$$\frac{d^2y}{dx^2} + k^2y = k^2(\delta + \varepsilon) \quad \text{with}$$

Boundary Conditions

$$x = o \quad y = 0, \quad \frac{dy}{dx} = 0$$

Figure 1-3 Lateral Bending Equation

If P/EI is defined as k^2 a linear second order differential equation for the displacement in terms of x is obtained. The solution to this equation will require two boundary conditions. With a built in base both the lateral deflection y and the slope of the member dy/dx will be zero at x =0.

Lateral Bending Analysis

The solution to the lateral bending equation is presented in Figure 1-4. It consists of two parts, a complimentary solution, y_c, and a particular integral, y_p. The complimentary solution satisfies the equation with the right side set equal to zero. The particular integral satisfies the right side of the equation. A classic functional form assumed for the complimentary solution for this linear second order total differential equation is $y_c = A \sin bx + B \cos bx$.

Substituting this into the differential equation in Figure 1-4 with the right side equal to zero can only be satisfied if b = k giving for the complimentary solution $y_c = A \sin kx + B \cos kx$. For the particular integral it is customary to assume a solution having the same form as the right side of the differential equation. Therefore y_p is chosen to be a constant C. Substituting this into the differential equation gives a value of $C = \delta + \varepsilon$. The total solution is the sum of the complimentary solution and the particular integral. The constants A and B must now be determined from the boundary conditions.

$$\frac{d^2y}{dx^2} + k^2y = k^2(\delta + \varepsilon)$$

Assume for the complementay solution

$$y_c = A\sin bx + B\cos bx \qquad \text{then}$$

$$-Ab^2\sin bx - Bb^2\cos kx + k^2A\sin bx + k^2B\cos bx = 0$$

so that $\quad b = k \quad$ and

$$y_c = A\sin kx + B\cos kx$$

Assume for the particular integral

$$y_p = C \qquad \text{then}$$

$$k^2C = k^2(\delta + \varepsilon) \Rightarrow C = (\delta + \varepsilon)$$

Final general solution

$$y = A\sin kx + B\cos kx + (\delta + \varepsilon)$$

Figure 1-4: Lateral Bending Analysis

Boundary Constraints

Since the member is built in at the base both its lateral deflection, y, and slope, dy/dx, must be zero at x = 0. These two conditions will be used to determine the constants A and B in the solution for the deflection as shown in Figure 1-5. Setting y = 0 at x = 0 results in B = - δ+ ε. Applying the second condition that the slope dy/dx = 0 at the base leads to A =0. The final equation defining the lateral deflection of this member is given by y = (δ + ε) (1- cos kx).

$$y = A\sin kx + B\cos kx + (\delta + \varepsilon)$$

$$\text{at } x = 0 \quad y = 0 \quad \text{then}$$

$$0 = B + (\delta + \varepsilon)$$

$$\text{so} \quad B = -(\delta + \varepsilon)$$

$$\text{at } x = 0 \quad \frac{dy}{dx} = 0$$

$$\text{with} \quad \frac{dy}{dx} = Ak\cos kx - Bk\sin kx \quad \text{then}$$

$$0 = A$$

and finally

$$y = (\delta + \varepsilon)(1 - \cos kx)$$

Figure 1-5: Boundary Conditions

Critical Buckling Load

Now consider that at the free end of the member the lateral deflection is delta. Substituting y = δ at x = L into the solution for the deflection results in δ = ε (1/coskL – 1) as illustrated in Figure 1-6. It is observed that if cos kL is set equal to zero δ becomes infinite. This is a singularity that can be used to determine kL and define a critical value of the compressive load P. If cos kL =0 then kL must be π/2. However, kL is also equal to the (√PEI)/L. If this value of P is designated as the critical load, P_{CR}, then $P_{CR} = \pi^2 EI / 4 L^2$.

$$y = (\delta + \varepsilon)(1 - \cos kx)$$

at x = L y = δ so that

$$\delta = (\delta + \varepsilon)(1 - \cos kL)$$

$$\delta - \delta \cos kL + \varepsilon - \varepsilon \cos kL$$

or $$\delta = \varepsilon\left(\frac{1}{\cos kL} - 1\right)$$

Now when coskl = 0 δ becomes infinite. Use this singular condition to determine kL and define a critical value of P

$$\text{if } \cos kL = 0 \text{ then } kL = \pi/2$$

but $$kL = \sqrt{P/EI}\, L, \text{ so that } \sqrt{P_{cr}/EI}\, L = \pi/2$$

or $$P_{cr} = \frac{\pi^2 EI}{4L^2}$$

Figure 1-6 Evaluation of Results

Re-examine Deflection

A physical interpretation of the critical load, P_{CR}, is best understood by looking at how the lateral deflection δ changes as the axial load P is increased. To examine this behavior the axial load P in the solution for δ from Figure 1-6 is first rewritten as δ = ε (sec kL -1), see Figure 1-7. Now the value of kL = (√P/EI) L is combined with the singular solution expression for P_{CR} to give kL = (√ P/P_{CR}) π/2. This is substituted into the rewritten equation for δ, resulting in a relationship that includes δ, ε, P and P_{CR} as shown in Figure 1-7 . This equation is now used to determine the behavior of δ with P as effected by ε and P_{CR}.

$$\delta = \varepsilon\left(\frac{1}{\cos kL} - 1\right)$$

or $$\delta = \varepsilon\left(\sec kL - 1\right)$$

now consider that

$$kL = \sqrt{P/_{EI}}\,L \text{ and introduce } \sqrt{P_{cr}/_{EI}}\,L = \pi/_{2}$$

to give $$kL = \sqrt{P/_{P_{cr}}}\,\pi/_{2}$$

so that finally

$$\delta = \varepsilon\left(\left[\sec\sqrt{P/_{P_{cr}}}\,\pi/_{2}\right] - 1\right)$$

Now plot P as a fuction of δ for different values of ε.

Figure 1-7 Deflection as a Function of P

Effect of Eccentric Loading

The behavior of the deflection δ as a function of the axial load P from the final equation in Figure1-7 is graphically illustrated in Figure 1-8 by a series of curves for different value of eccentricity, ε. For a value of ε = 0.01 the lateral deflection δ increases and approaches infinity asymptotically as the applied load P approaches P_{CR} but P_{CR} is never exceeded. As the value of the eccentricity decreases similar curves are obtained but the axial load P approaches P_{CR} at smaller values of lateral deflection δ. In the limit when the eccentricity is zero there is no lateral deflection of the member until P is equal to P_{CR} and then the lateral deflection is undefined. This behavior is the result of the mathematical singularity that was recognized in the solution of the deflection of the fixed and free end axially loaded member.

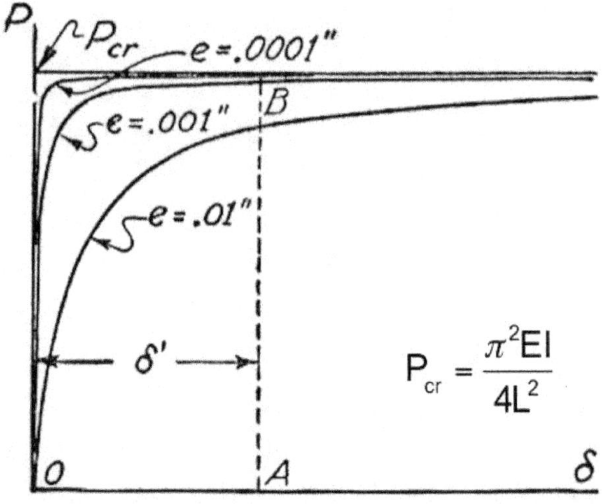

Figure 1-8: Effect of Eccentricity

The theoretical axial load that creates this undefined lateral instability behavior is referred to as the Euler Buckling Load and is used as the design basis of slender columns subjected to axial loading.

Chapter 2 – Effect of Support On Euler Buckling Loads

Introduction

Chapter 2 deals with determining how the support of the ends of the column affects the magnitude of the Euler Buckling load.

Classic End Supports

Four classic column support configurations are illustrated In Figure 2-1. The pinned–pinned column end support will only transmit collinear forces in line with the axis of the column. A moment and axial force at the built in end and a concentrated force at the pinned end restrain the pinned–fixed column. Moments and axial forces at both ends restrain the fixed–fixed column. The free-fixed column is only restrained at its base.

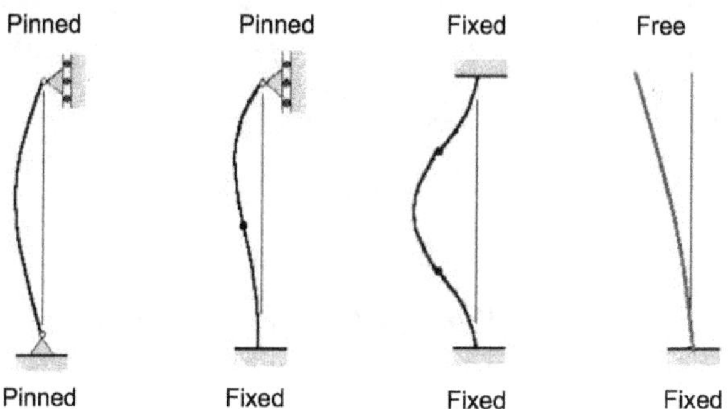

Figure 2-1 Classic End Supports

Pinned-Pinned Column

The pinned-pinned column can only support collinear axial forces P at its two ends. The bending moment as a function of x is simply M(x) = Py. Substituting this into the equation for lateral deflection due to bending and defining P over EI as k^2 gives the simple differential equation $d^2y/dx^2 + k^2y = 0$ as illustrated in Figure 2-2. The solution to this differential equation is subject to the two boundary conditions that at x = 0 the lateral deflection y = 0 and at x = L/2 the slope dy/dx must also be zero due to symmetry.

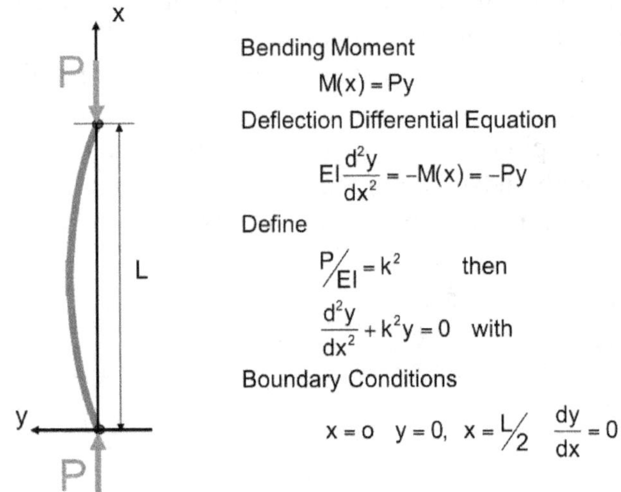

Figure 2-2 Pinned-Pinned Column

Solution to Differential Equation

The complementary solution to the differential equation is the same as assumed in Chapter 1, that is, $y_c = A \sin bx + B \cos bx$. Substituting this solution into the differential equation requires b to be equal to k as shown in Figure 2-3. Since the right side of the differential equation is zero there is no particular integral and y = A sin kx + B cos kx is the final general solution.

$$\frac{d^2y}{dx^2} + k^2y = 0 \quad \text{with}$$

Assume for the complementay solution

$$y_c = A\sin bx + B\cos bx \qquad \text{then}$$

$$-Ab^2\sin bx - Bb^2\cos kx + k^2A\sin bx + k^2B\cos bx = 0$$

so that $b = k$ and

$$y_c = A\sin kx + B\cos kx$$

For the particular integral

$$y_p = 0$$

Final general solution

$$y = A\sin kx + B\cos kx$$

Figure 2-3 Differential Equation Solution (Pinned-Pinned)

Buckling Load Calculation

Figure 2-4 illustrates how the constants A and B are determined using the physical condition that the lateral deflection is zero at x = 0 and from symmetry the slope, dy/dx, must be zero at x = L/2, the midpoint of the column. Applying the first condition gives A = 0. Condition 2 can only be satisfied if cos kL/2 = 0.

This implies that kL/2 must be π/2. Replacing k by √P/EI results in the critical bucking load, $P_{cr} = \pi^2EI/L^2$. As in Chapter 1 a singularity in the mathematical description of the bending under an axial load leads to a critical value of P that will cause lateral buckling instability.

$$y = A\sin kx + B\cos kx$$

at $x = 0$ $\quad y = 0$ $\quad$ then

$$0 = B$$

at $x = L/2$ $\quad \dfrac{dy}{dx} = 0$

with $\quad \dfrac{dy}{dx} = Ak\cos kx$

$$0 = Ak\cos k\,L/2$$

therefore

$$k\,L/2 = \pi/2 \quad \text{but since } k = \sqrt{P/EI}$$

then $\quad P_{cr} = \dfrac{\pi^2 EI}{L^2}$

Figure 2-4 Boundary Conditions (Pinned-Pinned)

Deflected Shape

Because the column is pinned at each end the deflection, y, at x = L must also be zero. Substituting this condition into the solution leads to the conclusion that the sin kl must be equal to π the same as the result already determined. Also recognizing that this can be used to define the value of k =π/ L the deflected shape of the buckled column can be written as y = A sin (x/L) as shown in Figure 2-5.

In other words the deflected shape of the column is a simple sine curve whose amplitude is undefined because of the singularity in the idealized mathematical solution that assumes the axial load acts through the centroid of the cross section with no eccentricity.

$$y = A\sin kx$$

but at $x = L \quad y = 0$ then

$$0 = A\sin kL$$

therefore $\sin kL$ must be zero so

$$kL = \pi$$

same as previous result

also

$$k = \pi/L$$

and the shape equation becomes

$$y = A\sin\pi\left(\frac{x}{L}\right)$$

Figure 2-5 Deflected Shape (Pinned-Pinned)

Pinned-Built In Member

A column with one end pinned and the other built in has the capability to withstand a resisting moment M_0 at the built in end as shown in Figure 2-6. In addition to the applied axial load P the pin end must also have the capability of resisting a horizontal component of force Q to satisfy equilibrium.

In this instance the bending moment at some position x on the column will consist of two components, one due to P and the other due to Q. Thus the bending moment is given by $M(x) = P\,y + Q\,(L-x)$. Substituting this into the differential equation for lateral bending results in a differential equation that is the same on the left as previously seen but with a term on the right involving a constant and the coordinate x (Figure 2-6). The boundary conditions at the fixed end are that the deflection and slope must both be zero. At the pinned end the deflection must be zero.

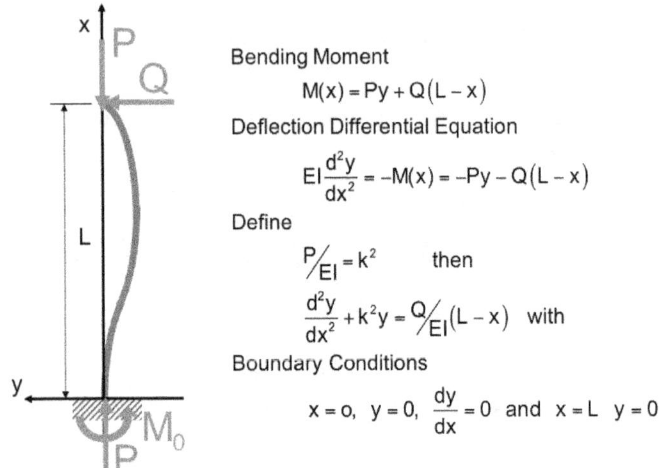

Figure 2-6 Pinned-Built In Member

Solution to Differential Equation

The complementary solution to the differential equation is the same as in the previous example, that is y_c = A sin kx + B cos kx. The particular integral y_p is assumed to be C (L - x). Solving for the constant C by substituting y_p in to the differential equation in Figure 2-7 gives C = Q /k^2EI.

The final general solution for the lateral deflection becomes y= A sin kx + B cos kx + (Q/ k^2EI) (L-x).

$$\frac{d^2y}{dx^2} + k^2y = \frac{Q}{EI}(L - x)$$

Assume for the complementay solution

$$y_c = A\sin bx + B\cos bx \qquad \text{then}$$

$$-Ab^2\sin bx - Bb^2\cos kx + k^2A\sin bx + k^2B\cos bx = 0$$

so that $\quad b = k \quad$ and

$$y_c = A\sin kx + B\cos kx$$

Assume for the particular integral

$$y_p = C(L - x) \qquad \text{then}$$

$$k^2C = \frac{Q}{EI} \;\Rightarrow\; C = \frac{Q}{k^2EI}$$

Final general solution

$$y = A\sin kx + B\cos kx + \frac{Q}{k^2EI}(L - x)$$

Figure 2-7 Solution to Differential Equation (Pinned-Built- In)

Buckling Load Calculation

Setting the deflection equal to zero at x=0 leads in Figure 2-8 to B =-Q(L/ k^2 EI). Applying the condition that the slope dy/dx = 0 at x = 0 results in A = - Q/k^3EI. The final equation for the lateral deflection becomes:
y = (QL/ EI) (cos kx – 1) – (Q/ k^3 EI) sin kx + (Q / EI) x.

One additional physical constraint must be satisfied. At the pinned end of the member the deflection must also be zero. Therefore at x = L, y =0.

$$y = A\sin kx + B\cos kx - \frac{Q}{k^2EI}(L - x)$$

at $x = 0$ $y = 0$ then

$$0 = B - \frac{QL}{k^2EI} \qquad B = \frac{QL}{k^2EI}$$

at $x = 0$ $\frac{dy}{dx} = 0$

with $\frac{dy}{dx} = Ak\cos kx - \frac{QL}{kEI}\sin kx + \frac{Q}{EI}$ then

$$0 = Ak + \frac{Q}{EI} \quad \Rightarrow \quad A = -\frac{Q}{kEI}$$

and finally

$$y = \frac{QL}{EI}(\cos kx - 1) - \frac{Q}{kEI}\sin kx + \frac{Q}{EI}x$$

Figure 2-8 Boundary Conditions (Pinned-Built-In)

Substituting this final boundary condition into the general solution for the lateral displacement results in the equation L cos kL - sin kL / k = 0 as shown in Figure 2-9.

$$y = \frac{QL}{EI}(\cos kx - 1) - \frac{Q}{kEI}\sin kx + \frac{Q}{EI}x$$

at $x = L$ $y = 0$ then

then $0 = \frac{QL}{EI}(\cos kL - 1) - \frac{Q}{kEI}\sin kL + \frac{Q}{EI}L$

or $0 = L\cos kL - \frac{\sin kL}{k}$

so that $\frac{\sin kL}{\cos kL} = \tan kL = kL$

solving gives $kL = 4.49$ (rad)

and $\sqrt{P_{cr}/EI}\, L = 4.49 \Rightarrow P_{cr} = \frac{20.2EI}{L^2} = \frac{2.04\pi^2 EI}{L^2}$

Figure 2-9 Additional Constraint (Pinned-Built In)

This can be further simplified to tan kL = kL. This transcendental equation must be solved numerically. By examining trigonometric tables for the tangent of an angle it is determined that when kl is equal to 4.49 radians then the tangent of kl is equal to kl. Substituting k equal to the square root of P_{cr} / EI for this singular value of k the Euler buckling load for a fixed – pinned column becomes $P_{cr} = 2.04\ \pi^2\ E\ I\ /\ L^2$.

Built-In Column

A column in compression with both ends built in is the last set of classical physical constraints to be analyzed. In this case there is a constraining moment M_o that resists angular rotation at the two ends of the member in addition to the compressive loads P.

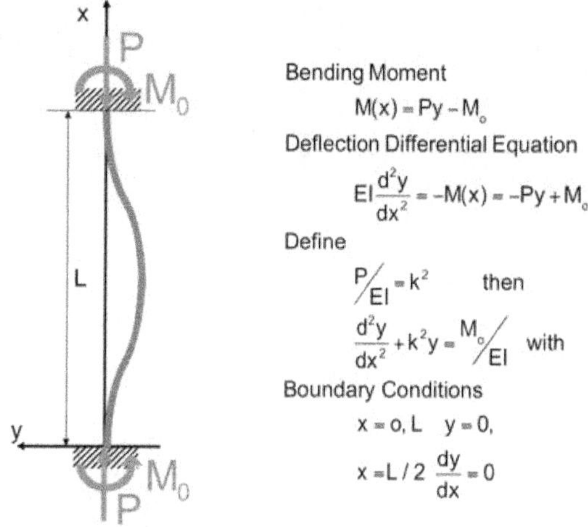

Figure 2-10 Built-In Column

In this configuration the bending moment as a function of x is given by M(x) = Py- M_o. Substituting this into the deferential equation for bending and again defining P/EI as k^2 gives a second order differential equation with a single constant term on the right hand side (Figure 2-10). Physical constraints of the support are described mathematically as the deflection y and slope is zero at x = 0 and L as well as the slope dy/dx = 0 at the center of the column where x = L / 2.

Solution to Differential Equation

The complementary solution to the differential equation is developed as for the previous two examples. This gives y_c = A sin kx + B cos kx. The particular integral will be a simple constant C = M_o / k^2 EI.

$$\frac{d^2y}{dx^2} + k^2 y = M_o / EI$$

Assume for the complementay solution

$$y_c = A \sin bx + B \cos bx \qquad \text{then}$$

$$-Ab^2 \sin bx - Bb^2 \cos kx + k^2 A \sin bx + k^2 B \cos bx = 0$$

so that $b = k$ and

$$y_c = A \sin kx + B \cos kx$$

Assume for the particular integral

$$y_p = C \qquad \text{then}$$

$$k^2 C = M_o / EI \Rightarrow C = \frac{M_0}{k^2 EI}$$

Final general solution

$$y = A \sin kx + B \cos kx + \frac{M_0}{k^2 EI}$$

Figure 2-11 Differential Equation Solution (Built-In)

Together this gives the final complete general solution for the lateral deflection in Figure 2-11 as:

$$y = A \sin kx + B \cos kx + M_o / k^2 EI.$$

Buckling Load Calculation

The boundary conditions corresponding to the physical constraints must now be applied. Setting the deflection equal to zero at x = 0 results in $B = M_o/k^2EI$ in Figure 2-12. Setting the slope dy/dx = 0 at x = 0 gives A =0. This gives the final result for the lateral deflection as $y = Mo/\ k^2EI\ (1 - \cos kx)$.

$$y = A\sin kx + B\cos kx + \frac{M_o}{k^2EI}$$

at $x = 0 \quad y = 0$ then

$$0 = B + \frac{M_o}{k^2EI} \qquad B = -\frac{M_o}{k^2EI}$$

at $x = 0 \quad \frac{dy}{dx} = 0$

with $\frac{dy}{dx} = Ak\cos kx + \frac{M_o}{kEI}\sin kx$ then

$$0 = A$$

and finally

$$y = \frac{M_o}{k^2EI}(1 - \cos kx)$$

Figure 2-12 Boundary Conditions (Built-In)

This solution must also satisfy the physical condition that at x = L / 2 the slope of the column dy/dx must be zero. This gives rise to the requirement in Figure 2-13 that $(M_o/\ kEI)\sin kL/2 = 0$. This will be true when $kL/2 = \pi$.

Hence, $k = 2\pi/L$ is a singular point resulting in the Euler buckling load for this configuration of

$$P_{cr} = 4\pi^2\ EI\ /\ L^2.$$

$$y = \frac{M_o}{k^2 EI}(1 - \cos kx)$$

at $\quad x = L/2 \quad \frac{dy}{dx} = 0$

$$\frac{dy}{dx} = \frac{M_o}{kEI}\sin kx$$

so that $\quad 0 = \frac{M_o}{kEI}\sin\frac{KL}{2}$

which requires that $\sin\frac{KL}{2} = 0$

or $\quad k\,L/2 = \pi$

hence $\quad k = \frac{2\pi}{L}$ and $P_{cr} = 4\frac{\pi^2 EI}{L^2}$

Figure 2-13 Additional Constraint (Built-In)

Deflected Shape

Because of the symmetric nature of the buckled deflected shape of the built in–built in column the deflection at x = L/2 can be designated as y_{max}. Substituting this condition along with k= 2π/L into the deflection equation results in a relationship between the maximum deflection and the restraining moment M_o at the built in ends. This gives y_{max} equal to $M_o L^2/2\pi^2 EI$ in Figure 2-14. Solving this relationship for M_o and substituting the result into the general solution for the lateral deflection gives $y = (y_{max} / 2)(1 - \cos 2\pi(x/L))$. In other words the deflected shape is a displaced cosine curve.

$$y = \frac{M_o}{k^2EI}(1-\cos kx)$$

at $x = L/2 \quad y = y_{max} \quad$ also $k = \frac{2\pi}{L}$

$$y_{max} = \frac{M_oL^2}{4\pi^2EI}(1-\cos\pi)$$

so that $\quad y_{max} = \frac{M_oL^2}{2\pi^2EI}$

therefore M_o is dependent on lateral displacement

Resubstituting for M_o in general deflection equation gives

$$y = \frac{y_{max}}{2}\left(1-\cos 2\pi\left(\frac{x}{L}\right)\right)$$

Figure 2-14 Deflected Shape (Built-In)

Generalized Results

The Euler critical buckling load for the four classically constrained columns subjected to compressive end loads can be generalized into a single expression as P_{cr} equal to a number "n" times the quantity (π^2 EI/ L^2)(see Figure 2-15). The number "n" takes on a specific value for each of the four constraint conditions considered.

For a free built in column the value is ¼. If the column is pinned at both ends the value of "n" is 1. For a pinned – built in column "n" becomes 2.04 and for a column built in at both ends the value of "n" is 4. In other words as the ends of the column become more constrained it takes an increasing value of the compressive force to produce buckling instability.

A column built in at both ends requires a compressive force 16 times the magnitude that will buckle a free - fixed column.

$$P_{cr} = n\frac{\pi^2 EI}{L^2}$$

where n depends on boundary conditions

$n = 1/4$	Free – Built in
$n = 1$	Pined – Pined
$n = 2.04$	Pined – Built in
$n = 4$	Built in – Built in

hence critcal buckling load for column built in at both ends is 16 times greater than if one end is free

Figure 2-15 Generalized Results

Chapter 3 –Material Property & Cross Sectional Geometry Effects

Introduction

Chapter 3 deals with the effect of material properties and cross section geometry on the buckling instability of columns together with criteria for design.

Alternate Form

To examine the effect of material properties and cross section geometry on buckling instability the critical Euler buckling load is rewritten in an alternate form. The area moment of inertia I of the cross section is replaced by the cross section area A times the radius of gyration "r" squared, see Figure 3-1.

Start with

$$P_{cr} = n\frac{\pi^2 EI}{L^2}$$

Recognize that I can be replaced by

$$I = Ar^2$$

where A = area of cros sec tion

r = radius of gyration

then $$P_{cr} = n\frac{\pi^2 EA}{\left(L/r\right)^2}$$

where $\left(L/r\right)$ is defined as the slenderness ratio.

Figure 3-1 Alternate Form

This permits the Euler critical buckling load to be expressed as $n(\pi^2EA)$ over the ratio $(L/r)^2$. The quantity (L/r) is referred to as the slenderness ratio.

Geometrically this is a measure of the "thinness" of the column compared to its length. E in the Euler buckling load is the modulus of elasticity of the column material.

Non-Dimensional Form

The critical buckling load P_{cr} is now divided by the modulus E times the area A to put the equation into dimensionless form. It is recognized that P_{cr} divided by the cross section area A is simply the magnitude of the compressive stress, σ_{cr}, required to produce buckling instability. The occurrence of the buckling phenomena is now expressed as $\sigma_{cr}/E = n\,\pi^2/(l/r)^2$ as shown in Figure 3-2

$$P_{cr} = n\frac{\pi^2 EA}{\left(L/r\right)^2}$$

Divide by EA

$$\frac{P_{cr}}{EA} = n\frac{\pi^2}{\left(L/r\right)^2}$$

Defne critical stress $\sigma_{cr} = \frac{P_{cr}}{A}$

so that finally

$$\frac{\sigma_{cr}}{E} = n\frac{\pi^2}{\left(L/r\right)^2} \qquad \frac{(psi)}{(psi)} = \frac{(in)^2}{(in)^2}$$

Figure 3-2 Non-Dimensional Form

Geometric Interpretation

Figure 3-3 is a graphic representation of the dimensionless critical buckling equation and its limitations. The compressive stress sigma, σ, due to axial loading of a column is plotted vertically against the slenderness ratio (L/r). The plotted line designated the Euler curve is a representation of the final equation developed in Figure 3-2. At low values of slenderness ratio the compressive stress required to initiate buckling instability is very high. As the slenderness ratio increases this required critical stress decreases dramatically and begins to level off. If the column material possesses a defined yield stress it is observed that for low values of the slenderness ratio the compressive stress in the column will reach this yield stress and produce permanent compressive deformation before the value of critical stress required to produce buckling instability is achieved. This effectively represents direct compressive failure of the column.

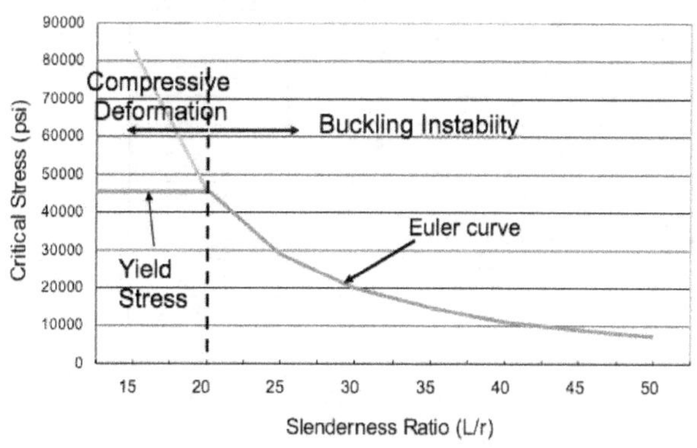

Figure 3-3 Stress vs. Slenderness Ratio

Where the yield stress value intersects the Euler curve a specific slenderness ratio is identified at which the behavior of the column goes from compressive failure to undefined lateral deflection initiated by buckling instability. Depending on the material properties of the column and its slenderness ratio the behavior of slender members due to axial loading undergo two separate modes of failure. This is frequently referred to as short column failure versus long column buckling.

Numerical Results

The table in Figure 3-4 lists numerical values for the non- dimensional critical stress as well as the critical buckling stress for free fixed steel and aluminum columns for slenderness ratios from 10 to 100. It is observed that for low values of slenderness ratio the critical stress required for buckling instability is

$$P_{cr} = \frac{\pi^2 EA}{4\left(L/r\right)^2} \Rightarrow \frac{\sigma_{cr}}{E} = \frac{\pi^2}{4\left(L/r\right)^2} \quad \text{(free - built in)}$$

(L/r)	$Sigma_{cr}/E$	Steel (psi)	Aluminum (psi)
0	Infinite	Infinite	Infinite
10	0.0246	739000	246000
20	0.0062	185000	62000
30	0.0027	82000	27000
40	0.0015	46000	15000
50	0.0010	29000	9800
60	0.0007	20000	6800
70	0.0005	15000	5000
80	0.0004	11000	3800
90	0.0003	9000	3000
100	0.0002	7300	2500

Figure 3-4 Numerical Results

significantly higher than the general level of yield stress for either material. At higher values of slenderness ratios the level of critical stress that will produce buckling instabilities are lower than general expected yield stresses of these materials. The difference in stress levels between the two materials for a given slenderness ratio is a consequence of the difference in their modulus of elasticity.

Graphic Representation

The numerical results listed in Figure 3-4 are plotted in Figure 3-5. This illustrates how the modulus and yield stress of the two materials affect the slenderness ratio at which transition occurs between short column and long column behavior. The upper curves are for a typical mild steel column with a yield stress of 47,000 psi. The lower curves are for an aluminum column with an assumed yield stress of 20,000 psi.

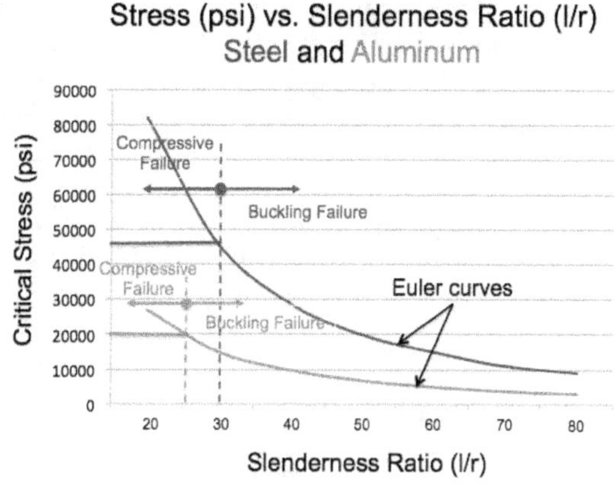

Figure 3-5 Graphical Representation

These two curves illustrate that the transition point from short to long column behavior can be different with the steel column acting like a short column for a higher slenderness ratio.

Effect of Eccentricity

In the analyses thus far, it has been assumed that the axial loads are applied at the centroid of the cross section so that there is no eccentricity affect. In reality this is virtually impossible to achieve. The question remains how does eccentricity of the loading effect the behavior of the column when material properties and cross section geometry are included. This question is answered by recognizing that the maximum compressive stress is made up of two terms, a stress that consists of a direct P/A compressive component and an Mc/I bending contribution, see Figure 3-6.

Maximum Bending Stress

$$\sigma_{max} = \frac{P}{A} + \frac{M_{max}\,c}{I}$$

for a free – fixed column

$$M_{max} = P(\varepsilon + \delta)$$

where

$$\delta = \varepsilon\left(\sec\frac{kL}{2} - 1\right)$$

with

$$k = \sqrt{\frac{P}{EI}} \quad \text{and} \quad I = Ar^2$$

Combining and rearranging gives

$$\sigma_{max} = \left(\frac{P}{A}\right)\left(1 + \frac{\varepsilon c}{r^2}\sec\frac{1}{2}\left(\frac{L}{r}\right)\sqrt{\frac{P}{EA}}\right)$$

Figure 3-6 Effect of Eccentricity

The maximum bending moment, $M_{max,}$ for a free built in column is P (ε +δ) from Figure 1-2. From Chapter 1 the free end deflection, δ, was found to be (sec(kL/2) -1). Substitutions for k = √(P/EI) and I =Ar^2 along with the deflection δ are now made in the stress equation. Combining and rearranging gives the final expression for σ_{max} in Figure 3-6. This permits the behavior of P/A as a function of the slenderness ratio to be determined.

Geometric Interpretation

For a specified modulus and an assumed yield stress of 16,000 psi plotting the results of the previous equation for P over A as a function of the slenderness ratio for selected values of ec/r^2 gives the set of curves presented in Figure 3-7. When the eccentricity is zero the two distinct modes of failure for short and long columns already described are observed with a flat yield stress curve up to a slenderness value of about 90 followed by the classical Euler curve. As the eccentricity factor is increased the value of P over A that will produce lateral deflection decreases and is represented by a continuous curve.

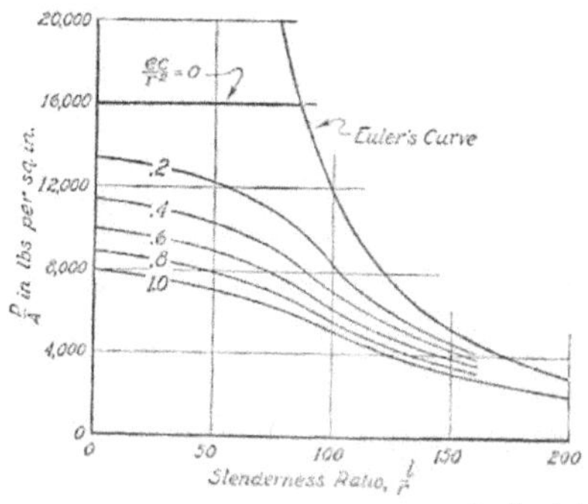

Figure 3-7 Geometric Interpretation

It is observed that the value of P/A never exceeds that predicted by the short or long column theory with no eccentricity. Hence these values of P/A with $\varepsilon=0$ represent an upper criteria limit for column design.

Design Correction

Real loaded columns never behave exactly as predicted by classical mathematical buckling instability theory. The region of greatest deviance is where the transition takes place between classic short column and long column theory. This is contributed to in large part by the effect of eccentricity. A variety of corrections have been proposed to help compensate for this unpredictable behavior by some reduction of the predicted classical load. One such correction inserts a parabolic behavior beginning at the yield stress and becoming tangent to the Euler curve at half the yield stress as illustrated in Figure 3-8. The reader is directed to other references dealing with design criteria for long columns for other proposed corrections. In all instances an appropriate factor of safety should be included to account for other unknown or indeterminate factors.

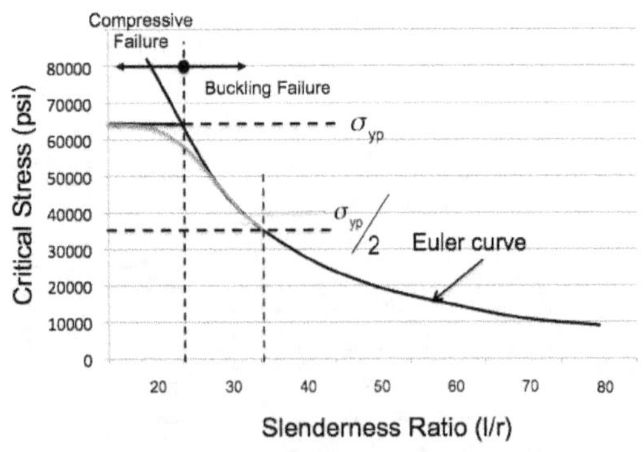

Figure 3-8 Design Correction

Chapter 4 – Shortening of Columns

Introduction

In this chapter, the relationship between the shortening of a column due to buckling instability and its maximum lateral displacement is determined. The mathematics are a little more difficult to follow but worth the effort in obtaining a useful final result.

Axial Displacement

When lateral displacement due to buckling takes place the ends of the column move closer together by some amount delta, δ, as illustrated in Figure 4-1. This axial displacement can be related to the maximum lateral displacement y_o. In terms of a coordinate "s" measured along the curve of the displaced column its length L can be expressed as the integral from 0 to L of ds. But ds can also be expressed as $\sqrt{(dx^2 + dy^2)}$ where x and y are a set of orthogonal coordinates measuring distance from the bottom of the column.

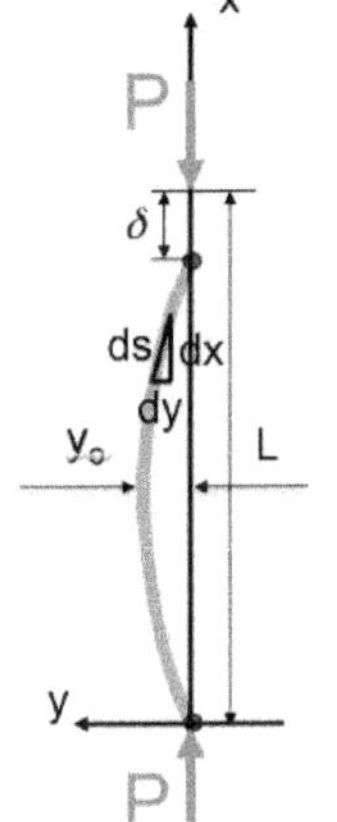

A column displaces an axial distance δ when it buckles under the load P_{cr} and can be related to its lateral displacement. Begin with

$$L = \int_0^L ds \qquad \text{but}$$

$$ds = \sqrt{dx^2 + dy^2} = dx\sqrt{1 + \left(\frac{dy}{dx}\right)^2}$$

$$\text{so} \qquad L = \int_0^L ds = \int_0^{L-\delta} \sqrt{1 + \left(\frac{dy}{dx}\right)^2}\, dx$$

Figure 4-1: Axial Displacement

Taking dx outside the square root gives ds = dx $\sqrt{(1+(dy/dx)^2)}$. Substituting this expression for ds in the first integral in Figure 4-1 results in L equal to the integral from 0 to (L-δ) of the quantity $(1+(dy/dx)^2)dx$.

Solution Simplification

The previous integral for the length L can be separated into two separate integrals as illustrated in Figure 4-2. By subtracting an integral from L-δ to L from an integral from 0 to L is equivalent to the original integral from 0 to L-δ. It is now recognized that the integral from L-δ to L of $(1+(dy/dx)^2)dx$ is simply δ because dy/dx is 0 over this portion of the integration. Therefore the integral for L from Figure 4-1 becomes L + δ equal to the integral from 0 to L of the quantity $\sqrt{(1+(dy/dx)^2)}$ integrated with respect to x.

but $$\int_0^{L-\delta}\sqrt{1+\left({}^{dy}\!/_{dx}\right)^2}\,dx=\int_0^{L}\sqrt{1+\left({}^{dy}\!/_{dx}\right)^2}\,dx-\int_{L-\delta}^{L}\sqrt{1+\left({}^{dy}\!/_{dx}\right)^2}\,dx$$

recognize that $\int_{L-\delta}^{L}\sqrt{1+\left({}^{dy}\!/_{dx}\right)^2}\,dx=\delta$ since $\frac{dy}{dx}=0$

then from previous page

$$L+\delta=\int_0^{L}\sqrt{1+\left({}^{dy}\!/_{dx}\right)^2}\,dx$$

since ${}^{dy}\!/_{dx}$ is small use series expansion, i.e.

$$\sqrt{1+\left({}^{dy}\!/_{dx}\right)^2}\cong 1+\frac{1}{2}\left(\frac{dy}{dx}\right)^2+\text{negligle terms}$$

Figure 4-2: Axial Displacement (contd.)

To simplify the integration the quantity $\sqrt{(1+(dy/dx)^2}$ is expanded into a series and approximated by just the first two terms neglecting all higher order terms.

Final Formulation

The quantity L +δ now becomes the integral from 0 to L of the quantity $(1+(dy/dx)^2/2)dx$. Integrating the first term in this integral is just the value L leaving finally that δ is equal to one half the integral from 0 to L of $(dy/dx)^2$ dx as shown in Figure 4-3.

Now consider the application of this relationship to a pinned-pinned column. In Chapter 2 the deflected shape for a pinned-pinned column was determined to be $y = y_o \sin \pi x/L$. This gives $dy/dx = y_o (\pi/L)(\cos \pi x/L)$. to be substituted into the integral defining delta.

Then

$$L + \delta = \int_0^L \sqrt{1 + \left(\frac{dy}{dx}\right)^2}\, dx = \int_0^L \left(1 + \frac{1}{2}\left(\frac{dy}{dx}\right)^2\right) dx$$

so that

$$\delta = \frac{1}{2}\int_0^L \left(\frac{dy}{dx}\right)^2 dx$$

For pined-pined column assume that

$$y = y_o \sin\frac{\pi x}{L}$$

$$\frac{dy}{dx} = y_o \frac{\pi}{L}\cos\frac{\pi x}{L}$$

Figure 4-3: Axial Displacement (contd.)

Specific Application

To perform the integration for delta for a pinned-pinned column a change of variables is first introduced. $\pi x/L$ is replaced by u which makes du equal to $\pi/L dx$ and the integral for delta becomes $y_o^2\pi^2/2L$ times the integral from zero to π of $(\cos u)^2 du$ in Figure 4-4. Performing this integration and evaluating the results between the limits of 0 to π gives the final result of $\delta = y_o^2 \pi^2 /4L$.

This gives

$$\delta = \frac{y_o^2\pi^2}{2L^2}\int_0^L\left(\cos\frac{\pi x}{L}\right)^2 dx$$

let $u = \frac{\pi x}{L}$ and $du = \frac{\pi}{L}dx$ and

$$\delta = \frac{y_o^2\pi}{2L}\int_0^\pi(\cos u)^2 du$$

$$\delta = \frac{y_o^2\pi}{2L}\left[\frac{u}{2}+\frac{1}{4}\sin 2u\right]_0^\pi$$

finally $\delta = \frac{y_o^2\pi^2}{4L}$ (pined-pined column)

Figure 4-4: Pinned-Pinned Column

Plastic Ruler Example

To demonstrate a numerical application of this result consider the buckling of a vertical thin plastic ruler that can be easily buckled to a finite lateral displacement y_o by simply pushing down on the top.

Numerical Result

The ruler is twelve inches in length, has a width of one inch and is about 1/16 of an inch thick. The modulus of the plastic is about 3.5 $\times 10^5$ psi. It will also be assumed that the maximum lateral buckled displacement is about 1 inch (see Figure 4-5).

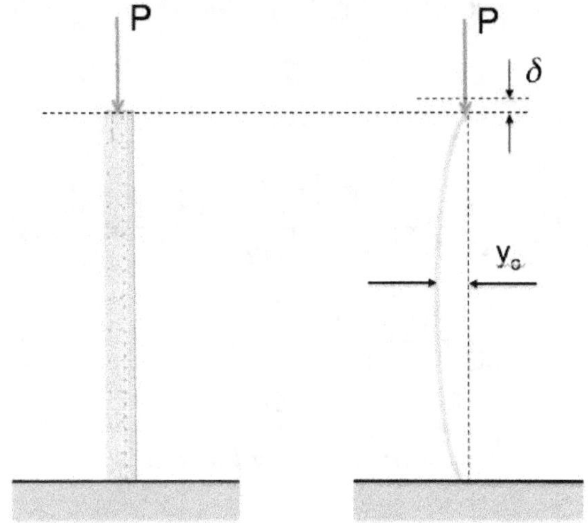

Figure 4-5: Thin Plastic Ruler

First the moment of inertia of the cross section given by the width, b, times the thickness cubed, h^3, divided by twelve is calculated to be 2.03 $\times 10^{-5}$ inches^4 (see Figure 4-6). Next P_{cr} is determined to be about 1/2 lb. that can be easily applied physically to create the buckled shape with a lateral deflection of 1 inch. Finally the axial downward displacement of the top of the ruler corresponding to this lateral displacement is calculated to be 0.20 inch, which appears very reasonable. The viewer can conduct an experiment with a plastic rule to confirm the results.

Bucking of flat plastic ruler :

$$L = 12 \text{ in.} \quad b = 1 \text{ in.}, h = \tfrac{1}{16} \text{ in.}$$

$$E = 3.5 \times 10^5 \text{ lbs/in}^2, \quad y_o = 1 \text{ in}$$

then

$$I = \frac{bh^3}{12} = \frac{1}{12}\left(\frac{1}{16}\right)^3 = 2.03 \times 10^{-5} \text{ in}^4$$

and

$$P_{cr} = \frac{\pi^2 EI}{L^2} = \frac{(3.14)^2 (3.5 \times 10^5)(2.03 \times 10^{-5})}{(12)^2} \cong 0.5 \text{ lb}$$

also

$$\delta = \frac{y_o^2 \pi^2}{4L} = \frac{(1)^2 (3.14)^2}{4 \times 12}$$

$$\delta = 0.20 \text{ in.} \quad \text{(reasonable)}$$

Figure 4-6: Numerical Result

The importance of being able to determine vertical movement in a buckled column for a specified lateral displacement is illustrated in the redesign problem described in detail in Chapter 5.

Chapter 5 – Thermostat Redesign Application

Introduction

This chapter deals with the performance analysis and redesign of a temperature activated switch whose behavior depends on the lateral instability of a slender metal strip. This design analysis illustrates the application of the buckling behavior topics and developments covered in the previous chapters.

Thermostat description

Shown in Figure 5-1 is a schematic illustration of the physical switch. The major components are three slender metal strips. The two outer strips are steel with cross section dimensions of 1/16" x 1/8". The center strip is aluminum with a cross section of 1/16" x 1/4". The strips are fastened together at each end by a single

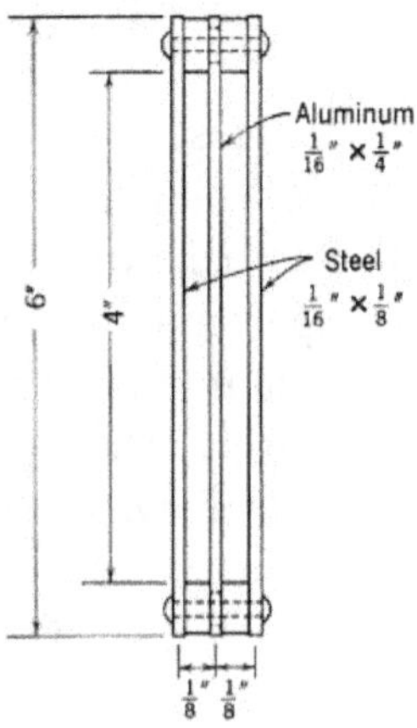

Figure 5-1 Thermostat Device

rivet through Bakelite spacers. This provides a gap of 1/8" between the centerlines of the two steel and aluminum strips. The overall length of the device is 6 in. while the free length of the strips is 4 in.

Design Specification

The switch is designed to close an electric circuit in a low cost product with a temperature increase of about 90° C. At the actuating temperature the central strip becomes laterally unstable and deflects sideways making contact with one of the outer strips closing the circuit.

It is desired to modify the design so that closing will occur with a rise of 20° C. If possible the external dimensions are to be preserved so the two switches will be interchangeable in the assembly of the product. It is also desirable to vary only the thickness of the strips if possible so their forming dies don't have to be changed.

Operational Performance

As the temperature increases above that at which the device was assembled both the steel and aluminum strips will elongate due to thermal expansion. With the thermal coefficient of expansion being greater for aluminum than for steel the aluminum strip would freely elongate more than the steel strips. However, the ends of the strips are attached together which results in the aluminum strip being put in compression while the steel strips are placed in tension to satisfy the physical restriction of the fixed ends.

This condition of geometric compatibility requires that the free thermal elongation of the aluminum strip minus its compressive contraction must be equal to the free thermal expansion of the steel strips plus their elongation due the tension that balances the compression in the aluminum strip. This is illustrated graphically in Figure 5-2.

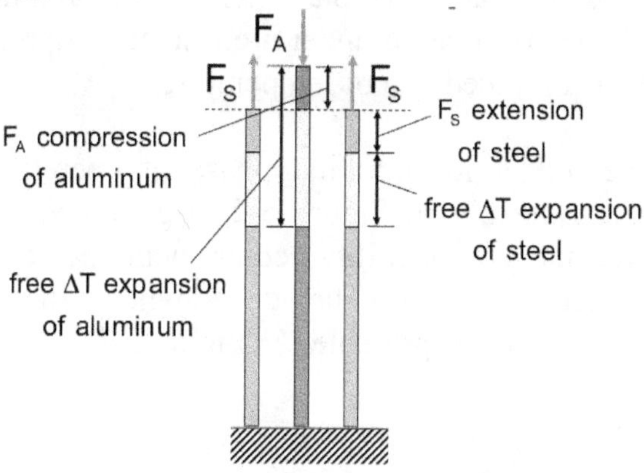

Figure 5-2 Geometric Compatibility

As the temperature continues to rise the compressive force in the aluminum strip will continue to increase. It will eventually reach the Euler critical buckling load for this slender member. From this point on the compressive force will remain constant even though the temperature continues to rise. Additional differential expansion of the aluminum and steel strips results in a lateral displacement of the aluminum strip due to buckling instability under this constant compressive load. The lateral deflection of the aluminum strip will continue to increase as the temperature rises further until contact is made with one of the steel strips.

Thermal Buckling Analysis

The behavior of the switch is analyzed as two separate problems. The first problem is modeled to describe the phase of operation in which the compressive force in the aluminum strip increases up its critical buckling load. The second problem is modeled to describe how the lateral displacement of the aluminum strip continues to increase under a constant compressive force with a continued rise of temperature.

The model for the first phase of operation as presented graphically in Figure 5-2 is governed by an equilibrium and compatibility requirement that can be expressed analytically. The development of these analytic expressions is presented in Figure 5-3.

Force equilibrium from model

$$\sum F = 0$$

$$F_A - 2F_s = 0 \Rightarrow F_A = 2F_s \qquad (1)$$

Deformation compatability

$$\Delta T \, exp_A - F_A \, comp_A = \Delta T \, exp_s + F_S \, ext_s$$

$$\alpha_A L\Delta T - \varepsilon_A L = \alpha_S L\Delta T + \varepsilon_S L$$

but $\varepsilon = \dfrac{\sigma}{E}$ and $\sigma = \dfrac{F}{A} \Rightarrow \varepsilon = \dfrac{F}{AE}$

and compatibility becomes

$$\alpha_A L\Delta T - \frac{F_A}{A_A E_A} L = \alpha_S L\Delta T + \frac{F_S}{A_S E_S} L \qquad (2)$$

Figure 5-3 Equilibrium and Compatibility

The equilibrium condition simply requires that the internal forces balance each other. This results in the magnitude of the compressive force F_A equal to the magnitude of the tensile forces $2F_S$. This is expressed by equation (1).

The geometric compatibility condition requires the free thermal extension of the aluminum strip minus its shortening due to the internal generated compression to be equal to the free thermal expansion of the steel strips plus their extension due to the generated tension. The free thermal extensions are given by the thermal coefficient of expansion α times the change in temperature ΔT multiplied by the length L of the strip. The subscripts A and S represent aluminum and steel respectively.

The compression in the aluminum is given by the compressive strain ε_A times the length L and the extension in the steel is given by the tensile strain ε_S times its length L. The strain, ε, is now replaced by the stress, σ, divided by the modulus of elasticity E. The stress, σ, is further replaced by the force on each strip divided by its cross sectional area. This leads to equation (2) in Figure 5-3.

Equation (1) is now used to eliminate F_S from equation (2). The result as illustrated in Figure 5-4 can be solved for ΔT in terms of F_A the compressive force in the aluminum strip. This equation governs the relation ship between the increase in temperature and the internal generated compressive force during phase one of the device operation. By substituting the critical buckling load for the aluminum strip into this equation the temperature increase required to reach this condition can be calculated. It is assumed here that the aluminum strip is built in at both ends.

Combine (1) and (2) and eliminate F_S

$$\frac{F_A}{A_A E_A} + \frac{F_A}{2\,A_S E_S} = (\alpha_A - \alpha_S)\Delta T$$

It will be conveninet to solve this equation for ΔT to determine temperature increase at which $F_A = P_{cr}$ for aluminum

$$\Delta T = \frac{F_A}{(\alpha_A - \alpha_S)}\left(\frac{1}{A_A E_A} + \frac{1}{2 A_S E_S}\right)$$

where $\quad F_A = P_{cr} = \dfrac{4\pi^2 E_A I_A}{L^2}$ (critical load for built in column)

Figure 5-4 ΔT – F_A Relationship

Critical Load Calculation

The problem data is now be gathered to determine the rise in temperature that will be required to raise the compressive force to the theoretical critical buckling load for the aluminum strip.

Problem Data

Al $\quad E_A = 10.2\text{x}10^6$ psi $\qquad$ Steel $\quad E_S = 30\text{x}10^6$ psi

$\alpha_A = 13.1\text{x}10^{-6}\ {}^1\!/_{^\circ F}$ $\qquad$ $\alpha_s = 6.5\text{x}10^{-6}\ {}^1\!/_{^\circ F}$

$A_A = \frac{1}{16}\text{x}\frac{1}{4} = .0156\ \text{in}^2$ $\qquad$ $A_S = \frac{1}{16}\text{x}\frac{1}{8} = .0078\ \text{in}^2$

$$I_A = (bh^3)/12 = \left(\frac{1}{4}\right)\left(\frac{1}{16}\right)^3 \Big/ 12 = 5.0\text{x}10^{-6}\ \text{in}^4$$

Now assume L = 4 in (distance between supports) then

$$F_A = P_{cr} = \frac{4\pi^2 E_A I_A}{L^2} = \frac{4(3.14)^2(10.2)(5.0)}{(4)^2} = 125\ \text{lb}$$

Figure 5-5 Problem Data

Also included in Figure 5-5 is the numerical calculation of the critical buckling load of 125 lb. for the aluminum strip assuming it is built in at both ends and its length is 4 in.

Critical Temperature Calculation

The temperature rise required to increase the compressive force to the critical buckling value is calculated in Figure 5-6. An increase of 156 ^{0}F or 69 ^{0}C represents the point at which the strip becomes axially unstable (can not take any additional compressive load) but has not yet undergone any lateral displacement

Begin with

$$\Delta T_{cr} = \frac{F_A}{(\alpha_A - \alpha_S)}\left(\frac{1}{A_A E_A} + \frac{1}{2A_S E_S}\right)$$

substituting values

$$\Delta T_{cr} = \frac{125}{(13.1 - 6.5)10^{-6}}\left(\frac{1}{(.0156)(10.2\times10^{6})} + \frac{1}{2(.0078)(30\times10^{6})}\right)$$

$\Delta T_{cr} = 156$ °F

This is temperature rise required to produce critical bucking load

Figure 5-6 Critical Load Temperature Calculation

Lateral Deflection Analysis

In the second phase of operation the differential expansion of the strips due to an additional temperature increase gives rise to a new geometric compatibility requirement. Since the compressive force remains constant there will be no further increase in force-induced strain in any of the strips. However, their lengths will change as the temperature continues to rise due to continued thermal expansion. It is the difference in the thermal expansions between the steel and the aluminum that must now be accounted for. Figure 5-7 illustrates graphically what takes place after F_A reaches P_{CR} and the temperature continues to rise.

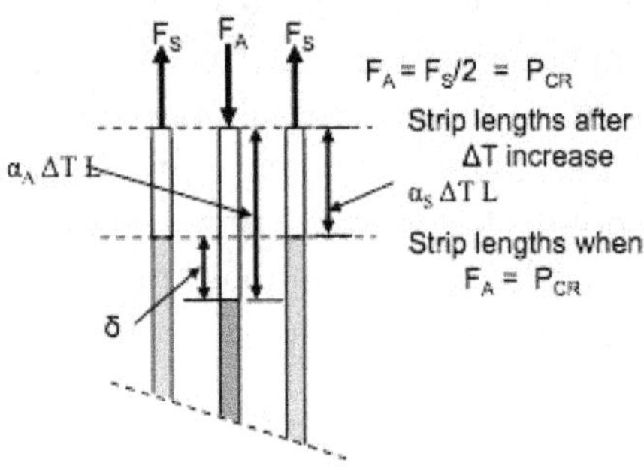

Figure 5-7 Post Buckling Geometric Compatibility

With the ends fastened together the Aluminum strip is actually pushed vertically down the distance δ from the position it had when the critical buckling load was reached.

$$\delta = \frac{1}{2}\int_0^L \left(\frac{dy}{dx}\right)^2 dx$$

For built in column

$$y = \frac{y_{max}}{2}\left(1 - \cos 2\pi\left(\frac{x}{L}\right)\right)$$

$$\frac{dy}{dx} = \frac{y_{max}}{L}\pi \sin 2\pi\left(\frac{x}{L}\right) \quad \text{so that}$$

$$\delta = \int_0^L \left(\frac{y_{max}}{L}\pi \sin 2\pi\left(\frac{x}{L}\right)\right)^2 dx = \frac{(y_{max})^2\pi^2}{4L}$$

Figure 5-8 y_{max}-δ Relationship

This effective compression results in a lateral displacement y_{max} that decreases the gap between the Aluminum strip and one of the steel strips. The general relationship between y_{max} and δ is covered in Chapter 4. For the present problem this is determined in Figure 5-8 using the shape function developed in Chapter 2 for a fixed-fixed column.

To determine the additional temperature rise to completely close the gap over that required to reach P_{CR} the distance δ is set equal to the difference in thermal coefficients of expansion time the column length L and the added temperature difference as shown in Figure 5-9.

Deflection Temperature Calculation

The required compression of the aluminum strip is calculated to be 2.4 x 10^{-3} in. This will completely close the gap y_{max} of 0.0625 in. This in turn results in a required additional temperature increase of 90°F or 32°C as determined in Figure 5-9.

$$\delta = (\alpha_A - \alpha_S) L \Delta T_a$$

or
$$\Delta T_a = \frac{\delta}{(\alpha_A - \alpha_S) L}$$

calculate δ and then ΔT_a

$$\delta = \frac{(y_{max})^2 \pi^2}{4L} = \frac{(.0625)^2 (3.14)^2}{4x4} = 2.4 x 10^{-3} \text{ in}$$

and
$$\Delta T_a = \frac{\delta}{(\alpha_A - \alpha_S) L} = \frac{(2.4 x 10^{-3})}{(13.1 - 6.5) x 10^{-6} (4)} = 90 \text{ °F}$$

Figure 5-9 Required Additional Temperature Increase

Total Temperature Rise

The total temperature increase required to raise the compressive force to P_{cr} in the aluminum strip plus the additional temperature increase required to close the gap is the sum of 156°F plus 90°F. This is equivalent to a temperature change of 137°C. This is significantly higher than the operational specification of 90°C. The question now is why? A reasonable answer is that the assumption of ideal fixed ends with a strip length of 4 in. is not appropriate for the single rivet fastening with Bakelite spacers that will possess some flexibility.

Effective Strip Length

To determine the effect of changing the length of the strips on the calculated results a length of 6 in. is used to recalculate the total temperature rise required. This recalculation gives a temperature change118°F or 65°C. A linear interpolation of these two results to obtain an effective length to account for the flexibility of the end restraint is a reasonable approach. This is illustrated in Figure5-10.

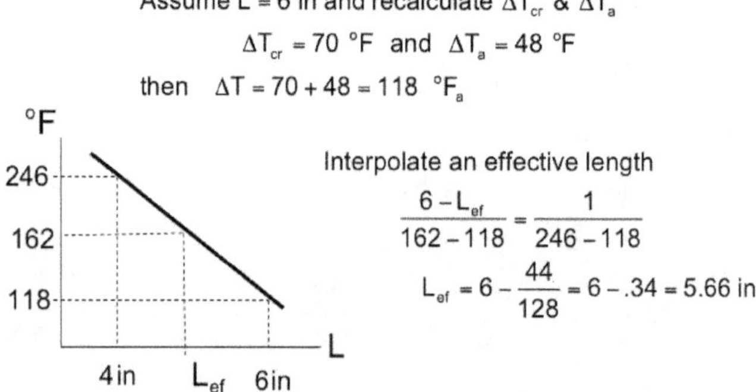

Figure 5-10 Determining an Effective Strip Length

Recalculating Required Temperature

Recalculating the required temperature increase with an effective strip length of 5.66 in. gives a result of 158°F or 87°C. This is reasonably close to the stated performance specification. This modified length will be used in the redesign to achieve an operational temperature chnage of 20°C or 36°F.

Redesign – Step 1

To begin a possible redesign all of the strips will simply be made thinner so that the aluminum strip will reach its critical buckling load at a lower temperature. As a trial dimension the new strip thicknesses will be reduced to 1/32 in. from 1/8 in. in the original design. The calculation of the reduced buckling load for this modification is shown in Figure 5- 11.

Change dimensions to obtain a total operating temperature of 20 °C, i.e.

$$\Delta T_d = 20\ ^\circ\mathrm{C} = \frac{9}{5}20 = 36\ ^\circ\mathrm{F}$$

Try reducing thicknesses of all strips to 1/32 in and adjust gap (y_{max}) to see if new performane can be acheived

New moment of inertia

$$I_A = \frac{bh^3}{12} = \frac{1}{4}\frac{1}{(32)^3}\frac{1}{12} = 6\times10^{-7}\ \mathrm{in}^4$$

Critical load

$$P_{cr} = \frac{4\pi^2 E_A I_A}{L^2} = \frac{(4)(3.14)^2(10.2\times10^{6})(6\times10^{-7})}{(5.66)^2} = 7.6\ \mathrm{lb}$$

Figure 5-11 Reduced Buckling Load

Redesign – Step 2

The temperature rise required to achieve this reduced critical load is calculated to be 18.8°F. This leaves 17.2°F to provide for sufficient lateral displacement to close the gap between the aluminum strip and one of the steel strips. This remaining temperature rise is used to determine the required gap size for closure in Figure 5-12.

Calculate required δ from

$$\Delta T_a = \frac{\delta}{(\alpha_A - \alpha_S)L}$$

or

$$\delta = (\alpha_A - \alpha_S)L\ \Delta T_a$$

$$\delta = (13.1 - 6.6)\text{x}10^{-6}(5.66)(17) = .23\text{x}10^{-3}\text{ in}$$

Now calculate y_{max} from

$$y_{max} = \frac{2}{\pi}\sqrt{\delta L} = \frac{2}{3.14}\sqrt{(.23\text{x}10^{-3})(5.63)} = 0.0367\text{ in}$$

This is larger than $1/32 = .0312$ inch and should work OK!

Figure 5-12 Required Gap Dimension

As observed the calculated required redesign gap dimension is 0.0367in. This is only slightly larger than 1/32 in. (0.312In.) making the redesign changes in dimensions feasible while retaining the overall dimensions of the device. The dies used to stamp out the strips could also be reused.

www.ingramcontent.com/pod-product-compliance
Lightning Source LLC
Chambersburg PA
CBHW071442180526
45170CB00001B/428

* 9 7 8 1 5 4 0 7 5 7 1 4 2 *